Methods in Molecular Biology

Series Editor
John M. Walker
School of Life and Medical Sciences
University of Hertfordshire
Hatfield, Hertfordshire, UK

For further volumes:
http://www.springer.com/series/7651

For over 35 years, biological scientists have come to rely on the research protocols and methodologies in the critically acclaimed Methods in Molecular Biology series. The series was the first to introduce the step-by-step protocols approach that has become the standard in all biomedical protocol publishing. Each protocol is provided in readily-reproducible step-by step fashion, opening with an introductory overview, a list of the materials and reagents needed to complete the experiment, and followed by a detailed procedure that is supported with a helpful notes section offering tips and tricks of the trade as well as troubleshooting advice. These hallmark features were introduced by series editor Dr. John Walker and constitute the key ingredient in each and every volume of the Methods in Molecular Biology series. Tested and trusted, comprehensive and reliable, all protocols from the series are indexed in PubMed.

Manuscript Instructions https://drive.google.com/file/d/1K4Az5vpAFVzYadJ3jKx2-QSo7Dqa_SNA/view?usp=sharing

RNA Labeling

Methods and Protocols

Edited by

Je-Hyun Yoon

Department of Oncology Science, University of Oklahoma, Oklahoma City, OK, USA

Humana Press

Editor
Je-Hyun Yoon
Department of Oncology Science
University of Oklahoma
Oklahoma City, OK, USA

ISSN 1064-3745 ISSN 1940-6029 (electronic)
Methods in Molecular Biology
ISBN 978-1-0716-5351-7 ISBN 978-1-0716-5352-4 (eBook)
https://doi.org/10.1007/978-1-0716-5352-4

This Humana imprint is published by the registered company Springer Science+Business Media, LLC, part of Springer Nature.
The registered company address is: 1 New York Plaza, New York, NY 10004, U.S.A.

Preface

Numerous studies on RNAs require a sophisticated method to label and detect RNAs of interest. Probing RNAs from the membranes originated from gel electrophoresis need radiotrophic or fluorescent labeling of complementary nucleotides. Direct labeling of RNAs is necessary for detecting and separating RNA–protein and RNA–RNA complexes. Moreover, RNA labeling via RNA metabolism also enables us to distinguish newly synthesized RNAs from old ones for subsequent labeling. RNA labeling is especially critical in research on RNA localizations, giving us spatial information on RNAs of interest in unique subcellular location.

Though methods of RNA labeling are widely utilized in a broader field of RNA biology, in the book on *RNA Labeling*, I aimed to provide a collection of comprehensive methods contributed by leading experts in RNA biology. The scope of this book ranges from classic methods of Northern blots to high-resolution microscopy as well as high-throughput RNA sequencing. It also covers in vitro methods of RNA labeling from tubes/membranes and in vivo methods of RNA detection by using the labeled complementary nucleotides.

The theme of this volume is very broad, and it was impossible to list all the details of individual methods. This book only covers RNA labeling methods frequently used in RNA biology of mammalian cells, but the same methods are commonly applicable to other model organisms, including Zebrafish, Drosophila, *C. elegans*, budding yeast, and prokaryotes, including *E. coli*. For researchers looking for RNA detection and Tagging, there are other books available, including one edited by Dr. Manfred Heinlein in volume 2166 and the other by Imre Gaspar in volume 1649.

Although I have been contributing to several chapters as a part of Methods in Molecular Biology, it is my first contribution to lead and edit a single book. I would like to appreciate all the contributing authors of individual chapters for their great expertise and efforts to share their valuable experiences to make their chapters sharper and brighter. It was also my great pleasure to work and communicate with the series editor, Dr. John Walker, and staff members of Methods in Molecular Biology to complete the book, RNA Labeling.

Je-Hyun Yoon
Department of Oncology Science
University of Oklahoma
Oklahoma City, OK, USA

Contents

Contributors

DANIEL BENHALEVY • *The Lab for Cellular RNA Biology, Shmunis School of Biomedicine and Cancer Research, George S Wise Faculty of Life Sciences, Tel Aviv University, Tel Aviv, Israel; The New Environmental School, Tel Aviv University, Tel Aviv, Israel*

SKYLER BRIGGS • *Department of Molecular Medicine, The Herbert Wertheim University of Florida, Scripps Institute for Biomedical Innovation and Technology, Jupiter, FL, USA; Department of Immunology and Microbiology, The Herbert Wertheim University of Florida, Scripps Institute for Biomedical Innovation and Technology, Jupiter, FL, USA; Skaggs Graduate School of Chemical and Biological Sciences, The Scripps Research Institute, Jupiter, FL, USA*

JAMES M. BURKE • *Department of Molecular Medicine, The Herbert Wertheim University of Florida, Scripps Institute for Biomedical Innovation and Technology, Jupiter, FL, USA; Department of Immunology and Microbiology, The Herbert Wertheim University of Florida, Scripps Institute for Biomedical Innovation and Technology, Jupiter, FL, USA*

SAKSHI DEWASTHALE • *Department of Biochemistry, Indian Institute of Science, Bangalore, India*

KYEONG-HO EUN • *Department of Anatomy and Cell Biology, Sungkyunkwan University School of Medicine, Suwon, Gyeonggi, South Korea*

BRIDGET FLOWERS • *Center for Genetic Diseases, Chicago Medical School, Rosalind Franklin University of Medicine and Science, North Chicago, IL, USA*

MICHELLE FONG • *Department of Medical Biochemistry and Cell Biology, Institute of Biomedicine, University of Gothenburg, Gothenburg, Sweden; Wallenberg Centre for Molecular and Translational Medicine, University of Gothenburg, Gothenburg, Sweden; Nucleic Acid Therapeutics Department, Discovery Sciences, AstraZeneca R&D, Gothenburg, Sweden*

YOUNGDAE GWON • *Department of Molecular Cell Biology, Sungkyunkwan University School of Medicine, Suwon, Republic of Korea; KIST-SKKU Brain Research Center, Sungkyunkwan University Institute for Convergence, Suwon, Republic of Korea*

JAEIL HAN • *Department of Microbiology and Molecular Biology, Chungnam National University, Daejeon, South Korea*

TILMAN HEISE • *Department of Pediatric Hematology, Oncology and Stem Cell Transplantation, University Hospital of Regensburg, Regensburg, Germany*

NOAH S. HELTON • *Department of Human Genetics, University of Michigan, Ann Arbor, MI, USA; Center for RNA Biomedicine, University of Michigan, Ann Arbor, MI, USA*

CHUL WOONG HO • *Department of Biological Sciences, College of Natural Sciences, Kangwon National University, Gangneung Campus, Gangneung, South Korea*

HANG T. HUYNH • *Department of Medical Biochemistry and Cell Biology, Institute of Biomedicine, University of Gothenburg, Gothenburg, Sweden; Wallenberg Centre for Molecular and Translational Medicine, University of Gothenburg, Gothenburg, Sweden*

MOONHYEON JEON • *Department of Chemical & Biomolecular Engineering, Korea Advanced Institute of Science and Technology, Daejeon, South Korea; Center for RNA Research, Institute for Basic Science, Seoul, South Korea*

HO-CHANG JEONG • *Department of Anatomy and Cell Biology, Sungkyunkwan University School of Medicine, Suwon, Gyeonggi, South Korea*

MYUNG HYUN JO • *Department of Life Sciences, Korea University, Seoul, Republic of Korea*

MAHEKDEEP KAUR • *The Lab for Cellular RNA Biology, Shmunis School of Biomedicine and Cancer Research, George S Wise Faculty of Life Sciences, Tel Aviv University, Tel Aviv, Israel*

MARK A. KAY • *Departments of Pediatrics and Genetics, Stanford University, Stanford, USA*

CHANKYUNG KIM • *Department of Life Science, Chung-Ang University, Seoul, Republic of Korea*

EUI TAE KIM • *Department of Microbiology and Immunology, College of Medicine, Jeju National University, Jeju, Republic of Korea; Jeju Research Center for Natural Medicine, Jeju National University Core Research Institute, Jeju, Republic of Korea*

HAK KYUN KIM • *Department of Life Science, Chung-Ang University, Seoul, Republic of Korea*

MINSEON KIM • *Department of Microbiology and Molecular Biology, Chungnam National University, Daejeon, South Korea*

SUJIN KIM • *Department of Life Sciences, Korea University, Seoul, Republic of Korea*

YOOSIK KIM • *Department of Chemical & Biomolecular Engineering, Korea Advanced Institute of Science and Technology, Daejeon, South Korea; Center for RNA Research, Institute for Basic Science, Seoul, South Korea*

ORLI KORCHEV • *The Lab for Cellular RNA Biology, Shmunis School of Biomedicine and Cancer Research, George S Wise Faculty of Life Sciences, Tel Aviv University, Tel Aviv, Israel*

HWI-HO LEE • *Department of Cancer Biomedical Sciences, Graduate School of Cancer Science and Policy, National Cancer Center, Goyang, South Korea*

JI WON LEE • *Department of Biological Sciences, College of Natural Sciences, Kangwon National University, Gangneung Campus, Gangneung, South Korea*

JONG-SUN LEE • *Center for Genetic Diseases, Chicago Medical School, Rosalind Franklin University of Medicine and Science, North Chicago, IL, USA; Department of Biochemistry and Molecular Biology, Chicago Medical School, Rosalind Franklin University of Medicine and Science, North Chicago, IL, USA*

JOO-HYUNG LEE • *Department of Biochemistry and Molecular Biology, McGovern Medical School, University of Texas Health Science Center, Houston, TX, USA*

JOONWOO LEE • *Department of Molecular Cell Biology, Sungkyunkwan University School of Medicine, Suwon, Republic of Korea*

JUYEON LEE • *Department of Molecular Biology, Pusan National University, Busan, Republic of Korea; Institute of Systems Biology, Pusan National University, Busan, Republic of Korea*

KYUNG-HA LEE • *Department of Molecular Biology, Pusan National University, Busan, Republic of Korea; Institute of Systems Biology, Pusan National University, Busan, Republic of Korea*

YONG SUN LEE • *Department of Cancer Biomedical Sciences, Graduate School of Cancer Science and Policy, National Cancer Center, Goyang, South Korea*

VIVIAN LOBO • *Department of Medical Biochemistry and Cell Biology, Institute of Biomedicine, University of Gothenburg, Gothenburg, Sweden; Wallenberg Centre for Molecular and Translational Medicine, University of Gothenburg, Gothenburg, Sweden*

KYUNG-WON MIN • *Department of Biological Sciences, College of Natural Sciences, Kangwon National University, Gangneung Campus, Gangneung, South Korea*

GAYATRI MOHANAN • *Department of Biochemistry, Indian Institute of Science, Bangalore, India*

STEPHANIE L. MOON • *Department of Human Genetics, University of Michigan, Ann Arbor, MI, USA; Center for RNA Biomedicine, University of Michigan, Ann Arbor, MI, USA*

VERONICA MOSKOVICZ • *The Lab for Cellular RNA Biology, Shmunis School of Biomedicine and Cancer Research, George S Wise Faculty of Life Sciences, Tel Aviv University, Tel Aviv, Israel*

IWONA NOWAK • *Department of Medical Biochemistry and Cell Biology, Institute of Biomedicine, University of Gothenburg, Gothenburg, Sweden; Wallenberg Centre for Molecular and Translational Medicine, University of Gothenburg, Gothenburg, Sweden*

HYUN JUNG OH • *Department of Biochemistry, College of Life Science and Biotechnology, Yonsei University, Seoul, South Korea*

PURUSHARTH I. RAJYAGURU • *Department of Biochemistry, Indian Institute of Science, Bangalore, India*

JEMIMA RIESS • *Department of Pediatric Hematology, Oncology and Stem Cell Transplantation, University Hospital of Regensburg, Regensburg, Germany*

AISHE A. SARSHAD • *Department of Medical Biochemistry and Cell Biology, Institute of Biomedicine, University of Gothenburg, Gothenburg, Sweden; Wallenberg Centre for Molecular and Translational Medicine, University of Gothenburg, Gothenburg, Sweden*

ENKHJIN SARUULDALAI • *Department of Cancer Biomedical Sciences, Graduate School of Cancer Science and Policy, National Cancer Center, Goyang, South Korea*

HYE-RIM SHIN • *Department of Biochemistry, College of Life Science and Biotechnology, Yonsei University, Seoul, South Korea*

GUNHILD SOMMER • *Department of Pediatric Hematology, Oncology and Stem Cell Transplantation, University Hospital of Regensburg, Regensburg, Germany*

SYLVIA WOOD • *Center for Genetic Diseases, Chicago Medical School, Rosalind Franklin University of Medicine and Science, North Chicago, IL, USA*

SHI HEON YOO • *Department of Cancer Biomedical Sciences, Graduate School of Cancer Science and Policy, National Cancer Center, Goyang, South Korea*

Part I

Classical Methods

Chapter 1

Labeling of Northern Blot Probes

Enkhjin Saruuldalai, Hwi-Ho Lee, Shi Heon Yoo and Yong Sun Lee

Abstract

Northern blotting (shortly "Northern") is a classical method for RNA detection and measurement but has been overshadowed by reverse transcription-polymerase chain reaction (RT-PCR) in recent days. This is largely due to the highly sensitive nature of qRT-PCR. Nonetheless, a unique advantage of Northern is its ability to distinguish between different RNA isoforms. Having this unique advantage, it is unlikely that Northern would become an obsolete method for RNA detection, especially in the field of molecular biology.

Keywords RNA, Northern blot, Probe, Isotope, Labeling

A gene can produce variable RNA products (termed "RNA isoforms") arising from initiation or termination of transcription at different sites. Additionally, a primary transcript is converted into a mature RNA(s) through several steps, including 5′- and/or 3′-trimming and splicing. All these events lead to RNA isoforms that share a common sequence but have different sizes. Since RNA is resolved on a gel and subjected to hybridization to a probe, a Northern blot can detect different RNA isoforms simultaneously using a single probe targeting the common sequence. Although qRT-PCR is also capable of producing a signal for all isoforms of RNA, given that the primer targets the common sequence, the signal itself is indistinguishable from one another. Therefore, qRT-PCR produces a readout that corresponds to the sum of isoforms sharing the common sequence. As mentioned earlier, the limitation of Northern, as compared to qRT-PCR, is its sensitivity. To overcome this, a Northern probe should be as strong as possible. The probe labeling step determines how well Northern detects an RNA isoform and measures RNA expression. In this chapter, we introduce Northern probe labeling methods in two parts: one for small RNAs ranging from 20–200

Enkhjin Saruuldalai, Hwi-Ho Lee, and Shi Heon Yoo contributed equally to this work.

Je-Hyun Yoon (ed.), *RNA Labeling*, Methods in Molecular Biology, vol. 3047,
https://doi.org/10.1007/978-1-0716-5352-4_1,

nucleotides (nts) in length and the other for large RNAs (>200 nts). Since we will focus on probe labeling and subsequent hybridization steps, earlier steps such as gel electrophoresis and transfer to membranes should be referred to in a book (Sambrook, J., & Russell, D. W. (2001). Molecular cloning: A laboratory manual (3rd ed.). Cold Spring Harbor Laboratory Press) or web-based resources.

Part 1: Small RNA Northern: Probe Preparation, Labeling, Hybridization, and Detection

1 Introduction

Since studies on small regulatory RNAs, with microRNAs (miRNAs) being the most prominent example, have burst onto the scene, Northern blotting for these small RNAs has been conducted regularly [1, 2]. Although PCR-based methods for small RNAs have been developed especially for scanty miRNAs, Northern is the most reliable method to ensure the correspondence of a signal to the actual existence of small RNAs, as highlighted in an early paper in the miRNA field [3].

In increasing the sensitivity of small RNA Northern, we highlight that the method of RNA isolation performed may significantly affect detection sensitivity. Some column-based methods may fail to recover small RNAs (<200 nts) efficiently, as shown in Fig. 1 (YSL, unpublished). Some column-based RNA isolation kits provide a workaround protocol for small RNAs, while other kits are specifically tailored for small RNAs. Therefore, researchers must choose a kit or protocol for RNA isolation cautiously.

The most commonly used probe is a short (18–24 nts) DNA oligonucleotide (oligo). There are other types of oligos, including RNA oligo and backbone-modified oligo such as locked-nucleic acids (LNA) [1, 4, 5], which have been utilized previously to produce satisfactory results. We advise researchers to carefully examine their own experimental parameters and determine the oligo most suitable as a different oligo may prove to be advantageous over DNA oligo in certain experimental conditions.

In this chapter, we will describe a protocol for 5′-end labeling of DNA oligo with γ-^{32}P-ATP. This protocol provides a specific and sufficient signal for well-known miRNAs. However, a modification of the protocol might be necessary for scanty small RNAs. For example, internal labeling (see the next section) could produce a stronger probe by allowing incorporation of multiple isotopic nucleotides in a single probe. Alternatively, RNA oligo and LNA might be considered as aforementioned.

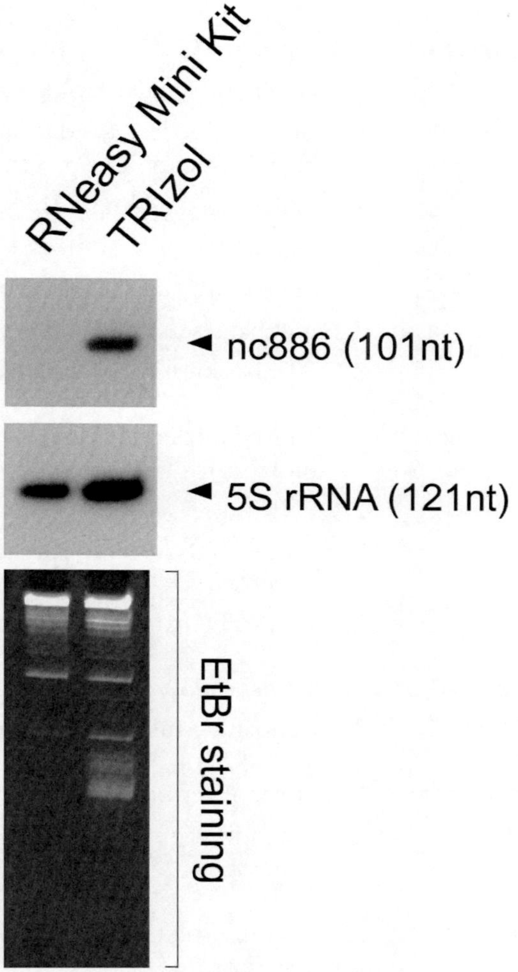

Fig. 1 Comparison of small RNA recovery between a column-based RNA extraction kit vs. TRIzol. Total RNA from an oral keratinocyte cell line, HOK-16B [6], was extracted by RNeasy Mini Kit (#74104, QIAGEN; Hilden, Germany) or the TRIzol reagent (#15596026, Invitrogen). The same amount of "total RNA" was loaded and resolved on a 15 % polyacrylamide gel. Top two panels, Northern of two small RNAs; bottom panel, ethidium bromide (EtBr) staining. RNeasy Mini Kit, a column-based method, yielded a significantly lower amount of the small RNAs than TRIzol. We clearly state that this result does not mean that RNeasy Mini Kit is inferior. RNeasy Mini Kit is optimized for large RNAs, and the recovery of large RNA portions would be better than TRIzol in the same amount of "total RNA"

2 Materials

2.1 Basic, Common Reagents or Equipment

1. Sephadex G-25 resin (exclusion limit ten base pair, fine).
2. Water (molecular biology grade).
3. Geiger-Mueller (GM) survey meter (model 3, Ludlum Measurements Inc.; Sweetwater, TX).
4. Radiation shielding lids.
5. Water bath or heat block (capable of 37 °C and 70–75 °C).
6. Microcentrifuge (capable of 12,000 × *g* or higher).
7. Incubators (rotator-equipped and shaker-equipped).
8. 50 ml conical tubes with leak-proof cap.

2.2 Probe Labeling

1. T4 Polynucleotide Kinase (PNK) (e.g., #M0201, New England Biolabs; Ipswich, MA).
2. PNK buffer (#B0201, New England Biolabs).
3. Probe (DNA oligo designed reverse complementary to a target).
4. γ-^{32}P-ATP (6000Ci/mmol, Revvity; Waltham, MA).

2.3 Sephadex G-25 Resin Preparation for Spin Columns

1. Sephadex G-25 resin (pre-swollen).
2. Centrifuge tube filters with 0.22 μm pore.

2.4 Pre-Hybridization and Hybridization of Northern Blot Membrane

1. Pre-transferred Northern blot membrane.
2. Hybridization buffer (ULTRAhyb®-Oligo Buffer, #AM8663, Invitrogen).

2.5 Membrane Washing

1. Washing solution: 2X SSC, 0.5% sodium dodecyl sulfate (SDS) (*see* **Note 1**).
2. Washing container with lid (leak-proof).

2.6 Detection of Signal

1. Whatman Chromatography Paper 3MM.
2. Cassette and X-ray film.
3. −80 °C freezer.

3 Methods

3.1 Preparation

3.1.1 Preparation of Sephadex G-25 Resin Suspension

1. Suspend Sephadex G-25 resin in DNase/RNase-free water at a 1:9 (w/v) ratio.
2. Swell the resin at room temperature for 3 h or overnight.

3. Degassing is recommended to prevent microbubble formation in the packed column. Apply a gentle vacuum to the resin suspension for > 30 m.
4. Adjust the suspension to 1:1 (v/v) ratio and store at 4 °C. The resin will be suitable for use for several months. If microbial contamination is a concern, add sodium azide (to 0.02–0.04% w/v).

3.1.2 Radiation Safety and Handling

Refer to general radiation safety instructions (e.g. https://www-pub.iaea.org/MTCD/Publications/PDF/PUB1781_web.pdf). In addition to strictly following them, we particularly highlight cautions below, based on our laboratory experience.

1. Before starting, survey with a GM survey meter to check radiation contamination in all equipment, handles, and benches along the workflow path.
2. Periodically monitor radiation on hands with the survey meter, to prevent secondary contamination to pipettes, tubes, centrifuge, refrigerators, etc.

3.2 Probe Labeling

1. Prepare the following reaction mix in 1.5 ml microcentrifuge tube and immediately tap to mix and spin down briefly (*see* **Note 2**).

Reagent	Volume
DNase/RNase-free water	5 µl
Oligo (5pmole/µl)	8 µl
10X PNK buffer	2.5 µl
γ-32P-ATP (6000 Ci/mmol)	8 µl
T4 PNK enzyme	1.5 µl
Total	25 µl

2. Incubate the mixture at 37 °C for 1 h (*see* **Note 3**).
3. To inactivate PNK, incubate the reaction tube in 70–75 °C water bath for 15 m.
4. Briefly spin down the tube to collect the mixture.

3.3 Sephadex G-25 Resin Preparation for Spin Columns

This step is to isolate the labeled DNA oligo from unincorporated γ-32P-ATP. Prepare Sephadex Spin Column during PNK inactivation incubation.

1. Fully pack swollen Sephadex resin into microcentrifuge spin columns by gentle pipetting.
2. Spin at 775 g for 3 m, discard flow-through.
3. Repeat steps 1 and 2 once more (total two spins).
4. Remove residual water in collection tubes with a pipette or suction.

3.4 Probe Elution

1. Carefully load the PNK-inactivated labeling reaction onto the center of Sephadex resin in a spin column.
2. Spin at 775 g for 3 m; collect eluate (20–25 μl).
3. Confirm the labeling efficiency by measuring radioactivity of the eluate (containing the labeled probe) and the remaining resin (containing unincorporated γ-32P-ATP). This can be formally done by measurement of radioactivity by a scintillation counter. However, we recommend a rough estimation with a survey meter in a proper setting (In the case of GM model 3, place the panhandle sensor (model #44-9) at ~15 cm distance at ×100 decade range.

   - If radioactivity in the remaining resin is significantly (~threefold) higher than the eluate, it indicates inefficient labeling and so all reagents and experimental steps must be checked.

3.5 Pre-Hybridization of Northern Blot Membrane

1. Roll the pre-transferred Northern membrane and place it inside a 50 ml conical tube.
2. Add 5 ml pre-warmed hybridization buffer gently (to avoid bubbles).
3. Cap the conical tube securely to prevent leakage
4. Ensure that the 5 ml buffer fully covers the membrane. If not, add more buffer. But be aware that more buffer results in less specific activity of the probe.
5. Place tube in rotating incubator set to a moderate speed (e.g., 15 rpm in Hybridization Incubator) for >30 m (up to several h) at 37 °C.

3.6 Hybridization with ^{32}P-Labeled Probes

1. Transfer all the eluted probe into the above 5 ml buffer in the pre-hybridization reaction.
2. Incubate at 37 °C overnight in a rotating incubator set to a moderate speed.
3. Confirm that tube caps are tight, and no leak occurs.

3.7 Membrane Washing

1. Pre-warm washing solution to 37 °C in a water bath and set the shaker incubator to 37 °C.
2. Place the hybridization reaction tube vertically in the rack, ensuring buffer pools at the bottom.
3. Transfer the membrane carefully into the washing solution in a flat-bottomed and leak-proof container. The membrane should stay flat, submerged in the washing solution.
4. Wash the membrane at 37 °C by shaking at moderate speed for 5 m and discard the washing solution in a designated radioactive waste container.
5. Repeat steps 3 and 4 again (total 2 washes).

6. Place a washed membrane in a 50 ml conical tube with ~25 ml washing solution at 37 °C for 30 m, in a rotating incubator set to a moderate speed.

3.8 Probe Detection

1. Remove excess washing solution from the membrane using 3MM paper.
2. Wrap the membrane flat with clear wrap and place it into a cassette.
3. Check radioactivity on the blot with GM survey meter to estimate exposure duration (*see* **Note 4**).
4. Expose the membrane to X-ray film at −80 °C.

Part 2: Large RNA Northern: Probe Preparation, Labeling, Hybridization, and Detection

4 Introduction

Compared to small RNAs, probe design is flexible for large RNAs (>200 nts). In principle, the length of the probe can be extended as long as the target RNA. DNA polymerization with α-^{32}P-dNTP on a template DNA allows multiple isotopic nucleotides to be incorporated in one probe molecule. This internal labeling will produce a stronger probe than 5′ (or 3′)-end labeling of only one isotopic nucleotide to a probe. However, internal labeling is not without its own problems. The template DNA, instead of the target RNA, can bind to the probe during the hybridization reaction. Usually, the template is a plasmid-driven DNA segment or a PCR product, both of which are double-stranded DNA (dsDNA). In this case, probes with both sense- and antisense-directions are generated, which might be a reason for off-target hybridization.

In this chapter, we will describe a protocol for internal labeling on dsDNA template with random priming. This is a simple way to produce a strong probe and can be modified according to experimental purposes and conditions. For example, synthetic single-stranded DNA as a template or strand-specific priming can be considered when off-target hybridization of an anti-sense RNA with the other probe strand is a concern. However, in most cases, this concern can be resolved by checking the existence of such an anti-sense RNA by using a sequence alignment tool, for example, Basic Local Alignment Search Tool (BLAST).

5 Materials

(For general equipment and common information, such as radiation safety, see part 1).

5.1 Random Primed DNA Probe Generation

1. *Bca*BEST™ Labeling Kit (#6046, TaKaRa; Kusatsu, Japan) This kit includes:
   - Random primer (9-mer).
   - 10X labeling buffer.
   - dNTP mixture (dATP, dGTP, and dTTP) (each 0.2 mM).
   - DNA polymerase (2 U/μl) (Cloned and modified *B. caldogenes* DNA polymerase).
2. dsDNA template (a PCR product or gel-isolated restriction fragment; 200~1000 nts long).
3. Water (molecular biology grade).
4. [α-^{32}P] dCTP (3000 Ci/mmol, 10 mCi/ml) (Revvity) (*see* **Note 5**).
5. EDTA (0.5 M).
6. Sephadex G-25 spin columns.

5.2 Pre-Hybridization and Hybridization with Probe

1. Pre-hybridization solution: 50% formamide (molecular biology grade), 3X SSC, 5X Denhardt's solution, 0.2% sodium pyrophosphate tetrabasic decahydrate, 1% SDS, 100 μg/ml DNA from herring sperm (hsDNA).
2. Hybridization solution: 50% formamide, 3X SSC, 5X dextran sulfate (molecular biology grade), 0.2% sodium pyrophosphate tetrabasic decahydrate, 1% SDS, 100 μg/ml hsDNA.

5.3 Washing

1. Wash solution I: 2X SSC.
2. Wash solution II: 0.1X SSC, 0.5% SDS.

6 Methods

6.1 Probe Labeling and Purification

1. Assemble the labeling reaction in a microcentrifuge tube (*see* **Note 6**).

Reagent	Volume
Template DNA	20–30 ng
Random primer	2 μl
Sterile purified water	x μl
Total	14 μl

2. Heat at 95 °C for 3 m and then cool on ice for 5 m.
3. Add 2.5 μl of 10× buffer, 2.5 μl of the dNTP mixture and 5 μl of [α-^{32}P] dCTP.
4. Add 1 μl of DNA Polymerase and incubate at 50–55 °C for 10 m (*see* **Note 7**).

5. Inactivate the enzyme by adding EDTA to a final concentration of 30 mm.
6. Heat at 95 °C for 3 m and then cool on ice.
7. Purify the radiolabeled probe from unincorporated nucleotides using a G-25 spin column (see part 1 for details).

6.2 Pre-Hybridization and Hybridization of Probe

1. Roll the pre-transferred Northern membrane and place it inside a 50 ml conical tube.
2. Add 5 ml pre-warmed hybridization buffer gently (to avoid bubbles).
3. Cap the conical tube securely to prevent leakage
4. Place the tube in the rotating incubator set to a moderate speed for 2 h at 42 °C.
5. Mix the probe and hsDNA (50 μl taken from 10 mg/ml stock solution, in the use case of 5 ml hybridization buffer, *see* **Note 8**) and denature them by placing it in a dry heating block at 95 °C for 10 m and then immediately chill on ice for 5 m.
6. Briefly spin down the probe and hsDNA mixture.
7. Add the denatured probe and the hsDNA mixture to the pre-warmed hybridization solution.
8. Incubate blot overnight at 42 °C in a rotating incubator set to a moderate speed.

6.3 Membrane Washing

1. Pre-warm washing buffer to 42 °C in the water bath and set the shaker incubator to 42 °C.
2. Pour the probe with the hybridization solution into a new 50 ml tube and store it at room temperature for future use.
3. Transfer the membrane carefully into 50 ml of wash solution I in a flat-bottomed and leak-proof container. The membrane should stay flat, submerged in the wash solution.
4. Wash the blot at room temperature by shaking at moderate speed for 5 m and discard the wash solution in a designated radioactive waste container.
5. Repeat steps 3 and 4 (total two washes).
6. Place the membrane into a 50 ml conical tube. Wash the blot at 42 °C in 25 ml of wash solution II for 30 m, in a rotating incubator set to a moderate speed.

6.4 Probe Detection

1. Remove excess buffer from the membrane using 3MM paper.
2. Wrap the membrane flat with clear wrap and place it into a cassette.
3. Expose membrane to X-ray film with cassette at −80 °C for 1–6 d.

7 Notes

1. 20X SCC buffer composition:
   - 3 M NaCl.
   - 0.3 M sodium citrate.
   - pH adjusted to 7.0 with HCl.
2. Do not reduce the DNA oligo amount even if the isotope amount is reduced.
3. Optional: incubation at a lower incubation temperature (35 °C recommended) can increase signal strength but may increase background; recommended for mRNA of low abundance.
4. Approximate exposure times based on survey meter (GM) readings described below:

Radiation level	Survey meter reading	Suggested exposure time*
Low	Detectable signal at ×0.1 decade range	24 h
Medium	Detectable signal at ×1 decade range	12 h
High	Detectable signal at ×10 decade range	3 h
Very high	Detectable signal at ×100 decade range	20 m

5. It is not necessary for the radiolabeled nucleotide to be [α-^{32}P] dCTP and can be changed to [α-^{32}P] dGTP, dTTP, or dATP depending on the sequence composition of the target RNA. However, when using *Bca*BEST™ Labeling Kit, because the provided dNTP mix does not contain dCTP, use of other radiolabeled nucleotides must be supplemented with a nonradioactive dNTP mix (excluding nucleotide of choice for radiolabeling).
6. The kit is designed to use 10 ng–1 μg of template DNA.
7. Incubation time: A probe with sufficient specific radioactivity can be obtained after 10 m of incubation. Incubation time can also be extended overnight without significant loss of activity. If a specific activity is low, overnight incubation may improve probe labeling efficiency.
8. The volume of hsDNA required varies depending on the volume of the pre-hyb/hybridization buffer and the concentration of stock hsDNA. Adjust the hsDNA volume to a final concentration of 100 μg/ml in the pre-hyb/hybridization buffer.

References

1. Valoczi A, Hornyik C, Varga N, Burgyan J, Kauppinen S, Havelda Z (2004) Sensitive and specific detection of microRNAs by northern blot analysis using LNA-modified oligonucleotide probes. Nucleic Acids Res 32(22):e175. https://doi.org/10.1093/nar/gnh171
2. Pall GS, Hamilton AJ (2008) Improved northern blot method for enhanced detection of small RNA. Nat Protoc 3(6):1077–1084. https://doi.org/10.1038/nprot.2008.67
3. Ambros V, Bartel B, Bartel DP, Burge CB, Carrington JC, Chen X, Dreyfuss G, Eddy SR, Griffiths-Jones S, Marshall M, Matzke M, Ruvkun G, Tuschl T (2003) A uniform system for microRNA annotation. RNA 9(3):277–279
4. Melton DA, Krieg PA, Rebagliati MR, Maniatis T, Zinn K, Green MR (1984) Efficient in vitro synthesis of biologically active RNA and RNA hybridization probes from plasmids containing a bacteriophage SP6 promoter. Nucleic Acids Res 12(18):7035–7056. https://doi.org/10.1093/nar/12.18.7035
5. Jorgensen S, Baker A, Moller S, Nielsen BS (2010) Robust one-day in situ hybridization protocol for detection of microRNAs in paraffin samples using LNA probes. Methods 52(4):375–381. https://doi.org/10.1016/j.ymeth.2010.07.002
6. Park NH, Min BM, Li SL, Huang MZ, Cherick HM, Doniger J (1991) Immortalization of normal human oral keratinocytes with type 16 human papillomavirus. Carcinogenesis 12(9):1627–1631. https://doi.org/10.1093/carcin/12.9.1627

Check for updates

Chapter 2

Complementary Approaches for Accurate Detection of tRNA-Derived Small RNAs (tsRNAs): Small RNA Northern Blotting and Dumbbell PCR

Chankyung Kim, Mark A. Kay and Hak Kyun Kim

Abstract

tRNA-derived small RNAs (tsRNAs) are generated by specific cleavage of transfer RNAs that are essential for delivering amino acids during protein synthesis. They have emerged as important regulators of gene expression across diverse biological contexts and have been implicated in multiple diseases, including cancers and neurological disorders. Advances in next-generation sequencing (NGS) have enabled high-throughput profiling of tsRNA abundance; however, extensive base modifications and heterogeneous termini frequently lead to inaccurate outcomes in small RNA sequencing. These technical limitations highlight the need for reliable validation strategies. Here, we present two complementary methods for tsRNA validation: small RNA northern blotting, which allows direct visualization of RNA size and abundance, and dumbbell PCR, which enables isoform-specific quantification with single-nucleotide resolution. Together, these approaches provide robust tools to validate tsRNA expression and facilitate mechanistic studies of their biological functions.

Keywords tRNA-derived small RNAs (tsRNA), Small RNA northern blot, Real-time PCR, Dumbbell PCR (Db-PCR)

1 Introduction

Transfer RNAs (tRNAs) are universally conserved molecules that play a central role in protein synthesis by delivering amino acids to ribosomes in accordance with codon recognition. Beyond this canonical function, tRNAs can undergo site-specific cleavage, generating a diverse group of smaller RNAs collectively referred to as tRNA-derived small RNAs (tsRNAs). These are also called tRNA-derived RNAs (tDRs) [1]. These fragments originate from distinct cleavage positions within pre-tRNAs or mature tRNAs, and their biogenesis and functions are largely determined by the cleavage position. For example, under cellular stress, angiogenin cleaves the anticodon loop of mature tRNAs to produce tRNA halves, which suppress overall

Je-Hyun Yoon (ed.), *RNA Labeling*, Methods in Molecular Biology, vol. 3047,
https://doi.org/10.1007/978-1-0716-5352-4_2,

translation and promote stress granule formation [2, 3]. Increasing evidence indicates that tsRNAs function as versatile regulators of gene expression, contributing to transcriptional silencing of nascent transcripts [4], post-transcriptional regulation through Argonaute-associated pathways [5, 6], and inhibition of translation by stress-induced fragments [3]. A notable example is the LeuCAG3′ tsRNA that has been shown to control ribosome biogenesis and support cancer cell growth [7, 8].

High-throughput small RNA sequencing has greatly facilitated the discovery of tsRNAs and enabled comprehensive profiling of their expression under diverse biological contexts and various diseases. However, small RNA sequencing method alone does not provide sufficient accuracy for tsRNA identification and profiling. Extensive chemical modifications in tsRNAs, together with heterogeneous termini such as 5′-P, 5′-OH, 3′-2′,3′-cyclic phosphate, or 3′-aminoacylated, can impede reverse transcription or reduce adapter ligation efficiency, which leads to sequencing library preparation method-dependent biases [9, 10]. These limitations require the need for sensitive and precise detection methods for tsRNAs.

Small RNA northern blotting provides direct, size-resolved visualization of tsRNA length and abundance and is largely insensitive to base modifications and terminal chemistries. Despite these advantages, the method has several important limitations. It often requires freshly prepared radioisotopes to achieve high sensitivity and demands much larger amounts of RNA input compared to PCR-based methods. Furthermore, it has inherently low throughput since targets are detected individually rather than simultaneously.

In contrast, dumbbell PCR (Db-PCR) employs stem–loop adapters that are ligated to both ends of the target RNA, generating a dumbbell-like molecule that can be selectively amplified by RT-PCR. With these adapters, Db-PCR can achieve single-nucleotide resolution and discriminate between highly similar tsRNA isoforms [11]. Nevertheless, since this method is affected by terminal chemistry and RNA modifications, additional pretreatments such as T4 PNK and demethylation are essentially required, which may in turn lead to biased or inaccurate quantification results.

In this chapter, we describe protocols for the detection of tsRNAs using both northern blotting and Db-PCR. Together, these complementary approaches provide reliable validation of tsRNA expression, enable clear distinction of isoforms, and facilitate mechanistic studies into the biological roles of tsRNAs in health and disease.

2 Materials

2.1 Sodium Arsenite Treatment and RNA Extraction

1. HeLa cell line
2. Dulbecco's Modified Eagle's Medium–high glucose
3. Fetal Bovine Serum

4. Antibiotic-Antimycotic (100×, Gibco™)
5. Sodium arsenite
6. 1× PBS
7. TRIzol™ Reagent (Invitrogen)
8. 1.7 mL nuclease-free microcentrifuge tube
9. Chloroform-isoamyl alcohol mixture (24:1)
10. Isopropanol
11. Ethanol
12. Nuclease-free water

2.2 Northern Blotting

2.2.1 PAGE and Transfer of RNA to Membrane

1. Total cellular RNA
2. Glass plate for gel (190 × 200 × 3 mm, thickness)
3. Spacer and comb, 0.8 mm thickness
4. RNaseZap (Thermo Fischer Scientific)
5. Nuclease-free water
6. 40% Acrylamide/Bis Solution, 19:1
7. 10× TBE
8. Urea
9. 10% (w/v) Ammonium persulfate (APS)
10. N,N,N′,N′-tetramethylethylenediamine (TEMED)
11. 2× RNA Gel Loading Dye
12. Whatman gel blotting paper, 0.8 mm thickness
13. Hybond-N+ nylon membrane
14. Trans-Blot® SD Semi-Dry Transfer Cell (Bio-rad)
15. UVP TL-2000 Ultraviolet Translinker (Artisan Technology group)

2.2.2 5′-End Labeling of Probe

1. DNA oligo for northern probe
   5′ Val-CAC probe:
   5′-GATAACCACTACACTACGGAA-3′
2. T4 Polynucleotide Kinase (10 U/μL) including 10× T4 Polynucleotide Kinase Buffer (Enzynomics)
3. ATP, [γ-^{32}P]–3000 Ci/mmol
4. MicroSpin G-25 Columns

2.2.3 Hybridization and Expose

1. PerfectHyb™ Plus Hybridization Buffer (Sigma-Aldrich)
2. Hybridization bottle w/Red PBT cap, 240 mL, Φ40 × h250 mm, thick 2.3 mm, GL45 screwcap
3. Hybridization incubator

4. Washing buffer (2× SSC, 0.1% SDS)
5. Storage phosphor screens and cassettes
6. Personal Molecular Imager™ (PMI™) System (Bio-rad)

2.3 Dumbbell PCR

2.3.1 Dbs Adapter Ligation

1. Total cellular RNA
2. 5'-Dbs adapter and 3'-Dbs adapter (*see* **Note 1**)
   Val (34nt) 5'-Dbs adapter:
   5'-ACGGAAACCGTCG/idSp/TGGAGTGTGTGCTTTGArCrG-3'
   Val (34nt) 3'-Dbs adapter:
   5'-/5Phos/CTCAGTGCAGGGTCCGAGGTATTCG-CACTGAGGAGGCG-3'
3. T4 Polynucleotide Kinase (10 U/μL) including 10× T4 Polynucleotide Kinase Buffer (Enzynomics)
4. 10 mM ATP
5. RiboLock RNase Inhibitor (Thermo Fischer Scientific)
6. Direct-zol RNA purification kit (Zymo)
7. 10× annealing buffer: 50 mM Tris-HCl (pH 8.0) and 100 mM $MgCl_2$
8. T4 RNA ligase 2 (Rnl2) (10 U/μL) including 10× T4 Rln2 reaction buffer

2.3.2 Reverse Transcription

1. RT primer: 5'-CTCAGTGCGAATACCTCGGACCCT-3'
2. SuperiorScript III Reverse Transcriptase (Enzynomics) including 5× First-Strand buffer
3. 0.1 M DTT
4. dNTP Mixture (10 mM each)
5. RNase H
6. Thermocycler

2.3.3 Real-Time PCR

1. Forward and Reverse primer for Dumbbell PCR
   Val (34nt) FW primer: 5'-GGAGTGTGTGCTTTGACGG-3'
   Val (34nt) RV primer: 5'-ACTGAGGAGGCGAACGTGA-3'
2. 2× TOPreal™ SYBR Green qPCR PreMIX (Enzynomics)
3. Real-Time PCR Thermal Cycler

3 Methods

3.1 Overview

Since Val-CAC tRNA halves are induced by sodium arsenite treatment, total RNAs were extracted from HeLa cells either treated with or without 0.5 mM sodium arsenite. For northern blotting, 10 μg of total RNA was resolved on a 15% denaturing gels, transferred to nylon membrane, UV-crosslinked, hybridized with a radioactive probe, washed, exposed to storage phosphor screen, and visualized using an imaging system (Fig. 1A). For dumbbell PCR, 1 μg of total RNA was ligated with 5′- and 3′-Dbs adaptors, followed by cDNA synthesis and real-time PCR (Fig. 1B).

3.2 Sodium Arsenite Treatment and RNA Extraction

All procedures should be performed on ice, and all solutions and materials must be RNase-free.

1. The day before the sodium arsenite treatment, plate HeLa cells to achieve ~70%–80% confluence on the day of treatment with DMEM medium supplemented with 10% FBS and 1× Antibiotic–Antimycotic. Incubate the cells at 37°C with 5% CO_2.
2. For sodium arsenite treatment, replace the medium with DMEM medium supplemented with 10% FBS and 1% Antibiotic–Antimycotic containing 0.5 mM sodium arsenite. Incubate the cells for 2 h at 37°C with 5% CO_2.
3. Aspirate the medium and gently wash the cells twice with chilled 1× PBS.
4. Add TRIzol reagent directly to the plate and pipette up and down to lyse completely. Transfer the lysates to a 1.7 mL microcentrifuge tube.
5. Add one-fifth volume of a chloroform-isoamyl alcohol mixture. Cap tightly and invert vigorously for 60 s until the solution becomes homogeneous.
6. Incubate the mixture for 2–3 min at room temperature to allow phase formation, then centrifuge at 18,000 *g* for 15 min at 4°C.
7. Carefully transfer the clear upper aqueous phase to a new 1.7 mL tube.
8. Add an equal volume of isopropanol to the transferred aqueous phase and mix by inverting.
9. Incubate the samples at room temperature for 10 min.
10. Centrifuge the mixture at 18,000 *g* for 15 min at 4°C.
11. Discard the supernatant without disturbing the pellet.
12. Add 900 μL of 70% ethanol and wash the pellet by brief vortexing.
13. Centrifuge the washed RNA at 18,000 *g* for 15 min at 4°C.
14. Completely remove the 70% ethanol without disturbing the pellet.

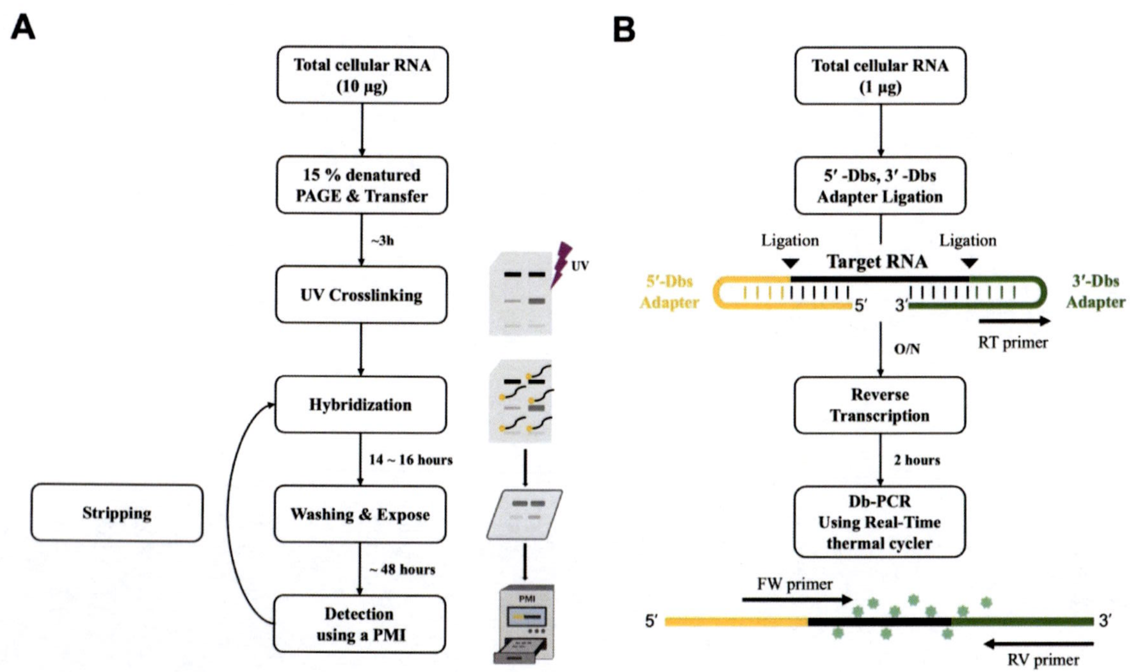

Fig. 1 Workflow of small RNA northern blot (**A**) and Dumbbell PCR (**B**)

15. Air-dry the RNA pellet for less than 10 min at room temperature.
16. Dissolve the RNA pellet in an appropriate volume of nuclease-free water and incubate for 7 min at 37°C to ensure complete solubilization.

3.3 Northern Blot

3.3.1 Gel Casting

1. Clean glass plates, spacers, and a comb with RNaseZap. Wipe all surfaces thoroughly, rinse with nuclease-free water, and allow to air dry.
2. Assemble the glass plates with spacers.
3. Prepare a 30 mL denaturing gel solution containing 15% acrylamide:bis (19:1), 1× TBE, and 8 M urea. Mix until the urea is completely dissolved and avoid bubbles.
4. Add 150 μL of 10% APS and 30 μL of TEMED to the denaturing gel solution from Step 3 and gently invert to mix.
5. Pour the denaturing gel solution from Step 4 into the assembled gel plates from Step 2 at a steady rate.
6. Insert the comb into gel solution from Step 5, leaving 0.5–1 mm gap above the glass plates.
7. Allow the gel to polymerize completely at room temperature.
8. After polymerization is complete, carefully remove the comb without disturbing the wells.
9. Rinse the wells of the gel from Step 8 with nuclease-free water.

3.3.2 Gel Running

1. Mount the denaturing gel in 1× TBE buffer and prerun at 400 V for at least 15 min.
2. Mix 10 μg of total RNA from Sect. 3.2 with an equal volume of 2× RNA Gel Loading Dye.
3. Heat the samples from Step 2 at 95°C for 5 min and immediately chill on ice for 3 min.
4. Rinse the wells with a running buffer using a syringe to remove residual urea.
5. Load the samples into the wells.
6. Run the gel at 400 V for 1.5–2 h in 1× TBE buffer until the bromophenol blue dye reaches 2–3 cm from the bottom of the gel.

3.3.3 Semidry Transfer

1. Prepare two dishes. Fill one with nuclease-free water, and fill the other with 0.5× TBE.
2. Carefully remove the gel from the glass plates.

3. Trim unused gel regions and place the gel in the 0.5× TBE to keep it fully wet.
4. Cut a positively charged nylon membrane and four sheets of blotting paper (0.8 mm thick) slightly larger than the gel (*see* **Note 2**).
5. Soak the membrane in nuclease-free water for 1 min, ensuring complete wetting.
6. Soak the membrane and blotting paper in 0.5× TBE, ensuring complete wetting.
7. Assemble the blotting stack in the following order (from bottom to top): two blotting papers, the membrane, the gel, and two additional blotting papers. Carefully remove air bubbles after each layer is added.
8. Wipe off excess buffer from the edges of the stack.
9. Transfer RNA from gel to membrane at constant power of 10 W (within a range of 6–12 W) for 45 min. The applied power should be adjusted according to the membrane size.

3.3.4 UV Crosslinking

1. Disassemble the transfer stack from Sect. 3.3.3, Step 9 and retrieve the membrane.
2. Gently rinse the membrane in 6× SSC with agitation.
3. Place the membrane in an Ultraviolet Translinker with the RNA-bound side facing up.
4. Crosslink the membrane at 120 mJ/cm^2 UV for 4 s with the crosslinker.

3.3.5 5′-End-Labeling of Probe

1. *In a 1.7 mL nuclease-free microcentrifuge tube, mix the following components:*
   - 2 μL of 20 μM antisense DNA oligo (5′ Val-CAC probe)
   - 2 μL of 10× polynucleotide kinase (PNK) buffer
   - 4 μL of [γ-32P]ATP (≥3000 Ci/mmol) (*see* **Note 3**)
   - 2 μL of T4 PNK (10 U/μL)
   - *Nuclease-free water to a final volume of 20* μL
2. Incubate at 37°C for 1 h.
3. Add 40 μL of nuclease-free water into the mixture from Step 2, then remove unincorporated [γ-32P] ATP from the probe using a MicroSpin G-25 column.
4. Denature the probe from Step 3 at 100°C for 1 min, then chill on ice for 3 min.

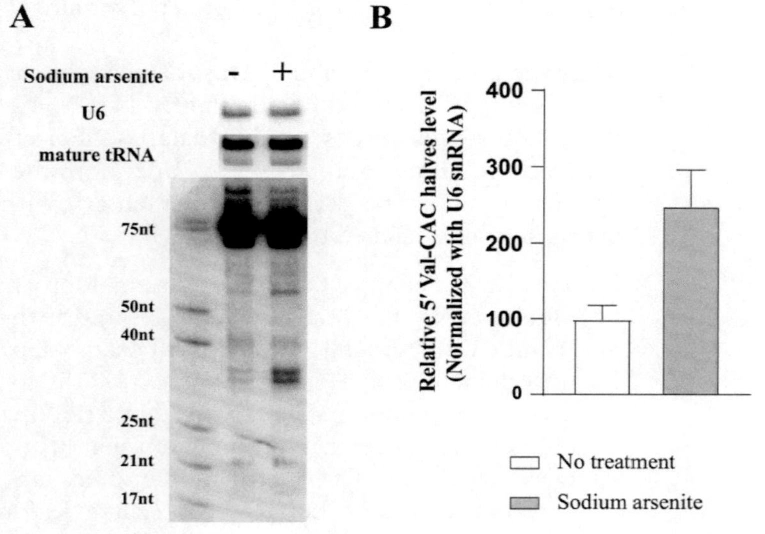

Fig. 2 Detection of the 5′Val-CAC tRNA halves (34 nt) by small RNA Northern blot and dumbbell PCR. Total RNA from HeLa cells either untreated or treated with sodium arsenite (0.5 mM, 2 h) was analyzed. (**A**) Northern blot using a DNA oligo probe complementary to the 5′Val-CAC tRNA halves (5′-GATAACCACTACACTACGGAA-3′). Bands between 25 and 40 nt were increased after arsenite treatment. (**B**) Db-PCR targeting the 34-nt 5′Val-CAC tRNA halves showed an increased signal after arsenite treatment, consistent with the northern blot results

3.3.6 Hybridization

1. Prehybridize the crosslinked membrane from Sect. 3.3.4 in PerfectHyb™ Plus Hybridization buffer at 34°C for 10 min with gentle rotation.
2. Add the purified probe (from Sect. 3.3.5 Step 4) to the hybridization buffer and hybridize at 34°C for 14–16 h with rotation.

3.3.7 Washing

1. After hybridization, leave the membrane in the hybridization bottle and discard the hybridization buffer.
2. Wash two times at 34°C for 10 min each with wash buffer (2× SSC, 0.1% SDS), followed by two additional washes at 36°C with the same buffer.

3.3.8 Expose & Detect

1. Wrap the washed membrane from Sect. 3.3.7 and place it in close contact with a phosphor screen for ~48 h (expose time can be adjusted depending on signal strength).
2. Detect the signals on the screen using a PMI (Fig. 2A) (*see* **Note 4**).

3.4 Dumbbell PCR

All procedures should be performed on ice, and all solutions must be RNase-free. Depending on the target tsRNA, pretreat samples prior to Sect. 3.4.1 adapter ligation step if required (*see* **Note 5**).

3.4.1 Dbs Adapter Design

Before designing the 5′- and 3′-Dbs adapters, the exact sequence of the target tsRNA should be determined based on your own small RNA sequencing data or a public database. For efficient ligation between the adapter and tsRNA by Rnl2, both the annealing and ligation sites must be perfect complementary, without any gaps, overlaps, or mismatched bases (Fig. 1B).

1. The 5′-Dbs adapter consists of a stem–loop and a 5′-annealing region that hybridizes to the 5′-end of the target RNA. Include an internal 1′,2′-dideoxyribose (/idSp/) in the loop to terminate reverse transcription across the loop and reduce unwanted secondary structure. Add two ribonucleotides at the 3′-end of the adapter to promote efficient ligation by T4 RNA ligase 2 (Rnl2). The length of the annealing region affects detection efficiency. Longer 5′-annealing region are generally associated with higher detection efficiency. In this study, an 8-nt 5′-annealing region was used to detect the 34-nt 5′Val-CAC tRNA halves.
2. The 3′-Dbs adapter consists of a stem–loop and a 3′-annealing region that hybridizes to the 3′-end of the target RNA. The adapter should contain a 5′-phosphate (/5Phos/) to enable ligation and a 5–7-nt 3′-annealing region that is perfectly complementary to 3′end of the target RNA. In this study, we used a 6-nt 3′-annealing region to detect the 34-nt 5′Val-CAC tRNA halves.
3. Analyze the adapter secondary structure with mFold [12] to avoid the formation of problematic stem-loop or self-complementary structures.

3.4.2 5′-Dbs Adapter Ligation

1. *In a 0.2 mL nuclease-free tube, combine 1* μg of total RNA from Sect. 3.2 with 20 pmol of 5′-Dbs adapter. Adjust final volume to 9 μL with nuclease-free water
2. *Denature at 90°C for 5 min and immediately add 1* μL of 10× annealing buffer to perform total 10 μL of reaction.
3. *Incubate at 37°C for 20 min to allow annealing and hybridization.*
4. *Add the following components to the same tube from Step 3:*
   - 2 μL of 10× T4 RNA ligase 2 Buffer
   - *0.1* μL of Rnl2 (10 U/μL) to provide 1 U
   - *7.9* μL of nuclease-free water
5. Incubate at 37°C for 30 min

3.4.3 3′-Dbs Adapter Ligation

1. *Add 1* μL of 3′-Dbs adapter (20 μM) to the ligation mixture from Sect. 3.4.2.
2. Incubate at 16°C for 30 min, followed by overnight incubation at 4°C.

3.4.4 Reverse Transcription

1. In a 0.2 mL PCR tube, prepare the following mixture:
   - 4 μL of ligated RNA (from Sect. 3.4.3)
   - 0.5 μL of dNTPs (10 mM each)
   - 1 μL of RT primer for target (5 μM)
   - 1 μL of RT primer for U6 snRNA as an internal control (5 μM)
   - 0.5 μL of nuclease-free water
2. Denature at 90°C for 2 min, then immediately chill on ice.
3. Add the following components to reach a total volume of 10 μ L:
   - 0.5 μL of 100 mM DTT
   - 2 μL of 5× First-Strand Buffer
   - 0.25 μL of RNase inhibitor (40 U/μL)
   - 0.25 μL of SuperiorScript III Reverse transcriptase (200 U/ μL)
4. Incubate at 55°C for 60 min, then 70°C for 15 min to stop reverse transcription.
5. Add 0.2 μL of RNase H (5 U/μL) and incubate at 37°C for 20 min, followed by 65°C for 20 min to inactivate the reaction.
6. Dilute the resulting cDNA from Step 5 to a 1:10 dilution with nuclease-free water.

3.4.5 Real-Time PCR

1. For each 12.5 μL, real-time PCR reaction, prepare the following:
   - 6.25 μL of 2× TOPreal™ SYBR Green qPCR PreMIX (*see* **Note 6**)
   - 1 μL of Forward primer (2.5 μM)
   - 1 μL of Reverse primer (2.5 μM)
   - 3 μL of 1:10 diluted cDNA from Sect. 3.4.4
   - 1.25 μL of nuclease-free water
2. Perform PCR procedure in a quantitative thermocycler with the following conditions:
   - Initial denaturation: 95°C for 12 min
   - 40 cycles of: 95°C for 10 s, 55°C for 15 s, 72°C for 20 s
   - Results were analyzed based on Ct values (Fig. 2B)

4 Notes

1. In the dumbbell adapter, underlined nucleotides indicate the single-stranded overhang base-pairing to the target RNA. /idSp/ denotes an internal dSpacer, and /5Phos/ indicates a 5′-phosphorylation.
2. During transfer, minimize the membrane size to just beyond the gel edges, as smaller membranes improve transfer efficiency.
3. Before working with the radioisotopes, confirm activity of the isotope stock using a scintillation counter, and proceed to 5′-end labeling of the probe only if the activity of radioisotopes is within the expected range.
4. When plan to probe multiple targets on the same membrane, strip the blot with stripping buffer (0.1× SSC, 0.5% SDS) at high temperature, then confirm that the previous signal has been completely removed before next hybridization.
5. RNA pretreatment may be required depending on the RNA sample. Treat PNK to generate 5′-phosphate and 3′-OH ends if the RNA carries a 5′-OH or other chemistry to block the ligation reaction. Perform demethylase treatment for tsRNAs containing methyl group that hinder base pairing or the reverse transcription step in Db-PCR.
6. If specificity is a concern, perform PCR with a TaqMan probe that spans the dumbbell junction instead of SYBR green method.

Acknowledgments

This research was supported by Suh Kyungbae Foundation (grant no. SUHF-22010038), the Korean Ministry of Science and ICT under Grant (grant nos. RS-2024-00358942 and 2022M3E5F1016861).

References

1. Holmes AD et al (2023) A standardized ontology for naming tRNA-derived RNAs based on molecular origin. Nat Methods 20(5):627–628. https://doi.org/10.1038/s41592-023-01813-2
2. Yamasaki S, Ivanov P, Hu G-F, Anderson P (2009) Angiogenin cleaves tRNA and promotes stress-induced translational repression. J Cell Biol 185(1):35–42
3. Ivanov P et al (2011) Angiogenin-induced tRNA fragments inhibit translation initiation. Mol Cell 43(4):613–623
4. Di Fazio A et al (2022) Dicer dependent tRNA derived small RNAs promote nascent RNA silencing. Nucleic Acids Res 50(3):1734–1752
5. Lee YS, Shibata Y, Malhotra A, Dutta A (2009) A novel class of small RNAs: tRNA-derived RNA fragments (tRFs). Genes Dev 23(22):2639–2649
6. Haussecker D et al (2010) Human tRNA-derived small RNAs in the global regulation of RNA silencing. RNA 16(4):673–695
7. Kim HK et al (2017) A transfer-RNA-derived small RNA regulates ribosome biogenesis. Nature 552(7683):57–62

8. Kim HK et al (2019) A tRNA-derived small RNA regulates ribosomal protein S28 protein levels after translation initiation in humans and mice. Cell Rep 29(12):3816–3824. https://doi.org/10.1016/j.celrep.2019.11.062
9. Cozen AE et al (2015) ARM-seq: AlkB-facilitated RNA methylation sequencing reveals a complex landscape of modified tRNA fragments. Nat Methods 12(9):879–884
10. Wang H et al (2021) CPA-seq reveals small ncRNAs with methylated nucleosides and diverse termini. Cell Discov 7(1):25
11. Honda S, Kirino Y (2015) Dumbbell-PCR: a method to quantify specific small RNA variants with a single nucleotide resolution at terminal sequences. Nucleic Acids Res 43(12):e77
12. Zuker M (2003) Mfold web server for nucleic acid folding and hybridization prediction. Nucleic Acids Res 31(13):3406–3415

Check for updates

Chapter 3

RNA Labeling for Measuring RNA Endonucleolytic Cleavage

Minseon Kim and Jaeil Han

Abstract

Although endonucleolytic cleavage of RNA is an essential process for maintaining cellular RNA homeostasis, its detection and quantification are often challenging. In this chapter, we present methods for detecting and quantifying RNA endonucleolytic cleavage events using Northern blot analysis. We describe two RNA labeling strategies for probe preparation: (1) a radiolabeled probe generated by 5′ end-labeling with [γ-^{32}P]ATP and polynucleotide kinase, and (2) a nonradioactive probe labeled with digoxigenin (DIG) or biotin for chemiluminescent detection. RNA samples are resolved by gel electrophoresis and transferred to a membrane. The labeled probes are then hybridized to the blotted RNA to detect specific cleavage products. These protocols enable both qualitative assessment of cleavage patterns and quantitative measurement of cleavage efficiency. We include step-by-step procedures from probe labeling to detection of target RNAs, as well as optional modifications to improve sensitivity and specificity. The methods are broadly applicable to analyzing RNA cleavage by various nucleases, providing a reliable combination of sensitivity for different experimental needs.

Keywords Northern blot, Endonucleolytic cleavage of RNA

1 Introduction

Endonucleolytic cleavage of RNA is a fundamental mechanism in RNA processing, turnover, and regulatory pathways [1, 2]. Many cellular RNases and endoribonucleases generate specific cleavage products from precursor or target RNAs, and measuring these cleavage events is crucial for understanding RNA maturation (e.g., rRNA or tRNA processing) and decay pathways (e.g., nonsense-mediated decay, no-go decay, and small RNA biogenesis) [3–7]. Northern blotting remains one of the most informative techniques to detect and map RNA cleavage events, as it separates RNA fragments by size and allows visualization of both precursor and cleaved products. Unlike RT-PCR-based methods, northern blots provide a direct qualitative view of fragment size distribution while still offering quantitative insight into RNA levels [8]. This is particularly valuable for confirming specific endonucleolytic cuts, as demonstrated in studies

Je-Hyun Yoon (ed.), *RNA Labeling*, Methods in Molecular Biology, vol. 3047,
https://doi.org/10.1007/978-1-0716-5352-4_3,

of pre-rRNA processing and mRNA decay where northern analysis delineated distinct cleavage intermediates [9] (Fig. 1).

A critical aspect of northern blot-based cleavage assays is the method of probe labeling used for detecting target RNA fragments. Historically, radiolabeling of probes have been the gold standard due to their high sensitivity and linear quantification range. In the classical approach, T4 polynucleotide kinase (PNK) is used to transfer the γ-phosphate of [γ-^{32}P]ATP to the 5′ end of an oligonucleotide probe, yielding a high-specific activity that can detect even low-abundance RNA fragments [10] (Fig. 2). However, radioactive methods pose safety hazards,

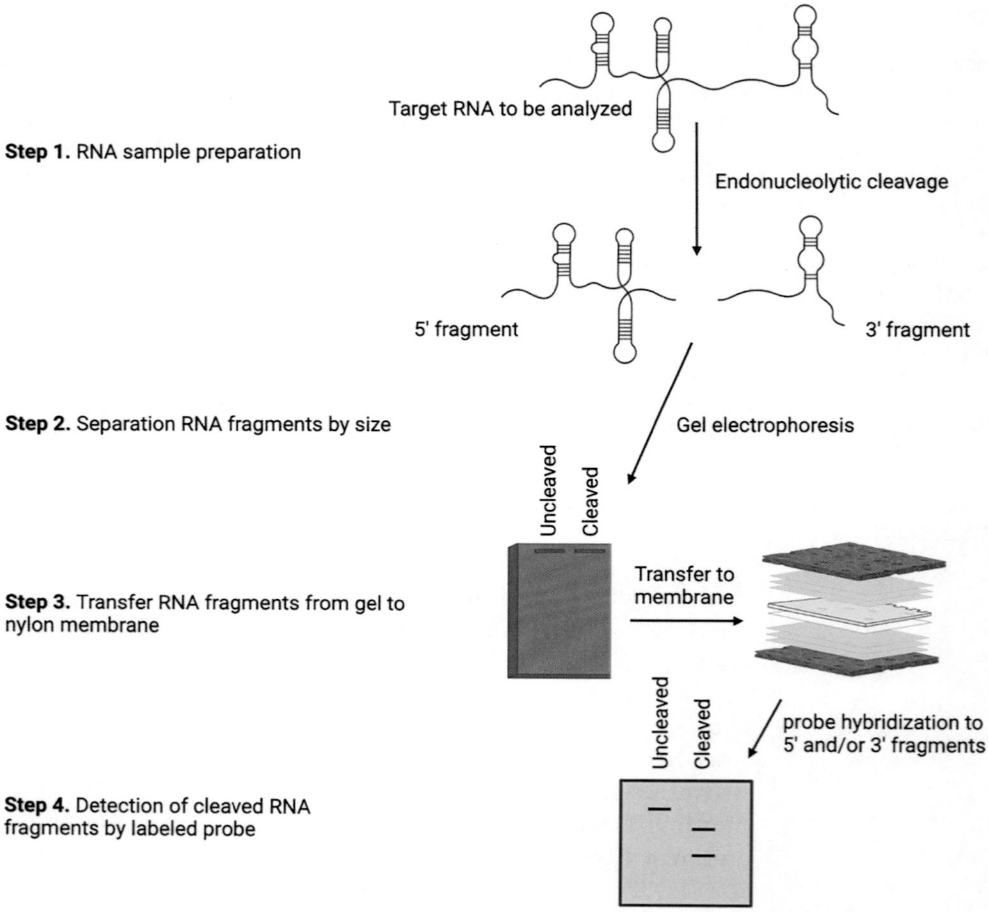

Fig. 1 Overview of labeling endonucleolytically cleaved RNAs by northern blot. Step 1: RNA sample preparation. RNA substrates are subjected to endonucleolytic cleavege by specific endoribonuclease or endogeneous cleavage events are directly analyzed by following steps using total RNAs. Step 2: RNA fragments are separated by gel electrophoresis. For relatively short RNAs (<500 nt), urea-polyacrylamide gel is suitable. For longer RNAs (>500 nt), agarose formaldehyde gel is recommended. Step 3. Separated RNAs are transferred to nylon membrane followed by cross-linking RNAs to the membrane. Step 4. Hybridization of labeled probes with the membrane will label cleaved and uncleaved target RNA, and following detection of the labeled RNA will reveal the cleavage events

require dedicated facilities, and radioactive probes have a short half-life and must be prepared fresh. As an alternative, nonisotopic labeling strategies have been developed using digoxigenin (DIG) or biotin conjugated to nucleic acids [11–13] (Fig. 2). DIG-labeled probes, detected via enzyme-linked chemiluminescence, have achieved sensitivity comparable to ^{32}P-based probes without the risks and regulatory burdens of radioisotopes [11]. With optimized probe design such as incorporation of locked nucleic acid bases and improved membrane-crosslinking techniques, nonradioactive methods can rival or even exceed the sensitivity of radiolabeling. DIG probes also offer practical advantages that experiments are more environmentally friendly and cost-effective over time.

In this chapter, we provide detailed protocols for measuring RNA endonucleolytic cleavage using northern blot analysis with either a radiolabeled probe or a nonradioactive DIG-labeled probe. We included steps that ensure both visualization of cleaved RNA fragments and quantitative measurement. In the Materials section, all necessary reagents and equipment are listed, including specialized reagents for each labeling approach. The Methods section outlines the procedure stepwise, from the sample preparation to gel electrophoresis, blotting, probe labeling, hybridization, washing, and signal detection. Where appropriate, we note alternatives and optimizations. Finally, in Notes, we discuss critical considerations and troubleshooting tips such as maintaining an RNase-free environment, optimizing probe design and hybridization conditions, and handling of radioactive (*see* **Notes 1–6**). By following these methods, researchers can sensitively detect specific RNA cleavage events and accurately quantify the extent of cleavage.

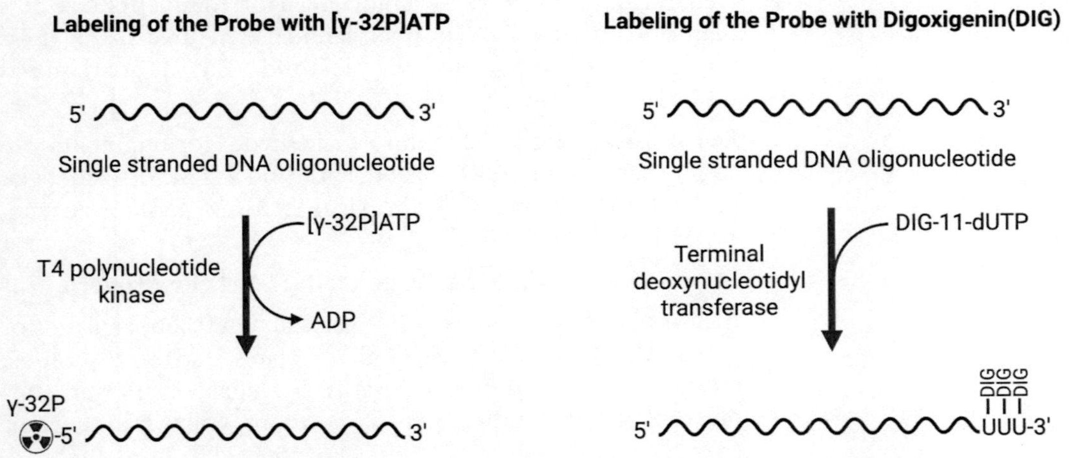

Fig. 2 Schematics of oligonucleotide probe labeling with either radioisotope or nonradioactive molecules. T4 polynucleotide kinase (PNK) adds radioactive ^{32}P from [γ-^{32}P]ATP to 5′ end of oligonucleotide. Radioactive probes can be detected by autoradiography. Digoxygenin (DIG)-dUTP can be added to the 3′ end of oligonucleotide by Terminal deoxynucleotidyl transferase (TdT). DIG-labeled probes can be detected using anti-DIG antibody conjugated to alkaline phosphatase (AP) by measuring enzymatic activities of AP

2 Materials

2.1 RNA Sample Preparation

1. Total RNA isolated from cells of interest, containing target RNAs to be analyzed. Typical yield should be sufficient to load 5–20 μg total RNA per lane for northern blot detection of moderately abundant transcripts.
2. Trizol.
3. DEPC (Diethyl Pyrocarbonate).
4. Chloroform.
5. RNaseZap (#R-2020, Sigma-Aldrich).
6. Ethanol.

2.2 In Vitro Cleavage Assay

1. The endonuclease enzyme to be tested. This may be a purified nuclease, an immunoprecipitated protein, a ribozyme, or a cellular extract if analyzing endogenous activities.
2. Reaction buffer: 20–50 mM Tris-HCl, pH 7.0–8.0, 50–150 mM NaCl or KCl, and required cofactors such as $MgCl_2$ or $CaCl_2$ depending on the nuclease used.
3. RNasin (#N2511, Promega).

2.3 Denaturing Gel Electrophoresis

1. 2× RNA loading dye: 50% formamide, 0.01% SDS, 0.01% bromophenol blue, 0.005% Xylene Cyanol, 0.5 mM EDTA.
2. ssRNA Ladder (#N0362S, NEB) for size range 500–9000 bases.
3. ssRNA ladder (#N0364S, NEB) for size range 50–1000 bases.
4. Denaturing urea-polyacrylamide gel (for fragments size less than 500 nt): 6%–18% polyacrylamide, 7 M urea in 1× TBE buffer. Ammonium persulfate (APS) and TEMED are required for polymerization.
5. Denaturing agarose gel with formaldehyde (for fragments size larger than 500 nt): 1%–1.2% Agarose and 2.2 M formaldehyde in 1× MOPS buffer (20 mM MOPS, 5 mM sodium acetate, 1 mM EDTA, pH 7.0).
6. Ethidium Bromide or SYBR gold (#S11494, Invitrogen).
7. Electrophoresis equipment: Vertical gel apparatus for polyacrylamide gels (with glass plates, spacers, combs, etc.) or horizontal gel apparatus for agarose. Power supply capable of running gels at constant voltage or current (5–8 V/cm).

2.4 Membrane Transfer

1. Positively charged nylon membrane for blotting RNA.
2. Transfer buffer: 1× TBE (90 mM Tris base, 90 mM Boric acid, 2 mM EDTA, pH 8.0) for urea-polyacrylamide gel

electrophoresis or 10× SSC (1.5 M NaCl, 0.15 M $Na_3C_6H_5O_7$) for formaldehyde agarose gel.

3. Filter paper, sponge pads, and a blotting apparatus.
4. For Electroblotting: A semidry transfer unit or tank transfer system. Recommended for polyacrylamide gels and small RNAs for efficient transfer in 0.5× TBE buffer.
5. For Capillary transfer: For formaldehyde agarose gels, upward capillary blotting onto a nylon membrane using 10× SSC transfer buffer is standard.

2.5 RNA Fixation on Membrane

1. UV crosslinker or transilluminator (254 nm) for fixing RNA to the membrane after transfer.
2. N-(3-Dimethylaminopropyl)-N′-ethylcarbodiimide hydrochloride (EDC) if using the EDC method for small RNAs (*see* **Note 7**).

2.6 Prehybridization

1. ULTRAhyb®-Oligo (#AM8663, Ambion) for small RNA northern blot.
2. Standard northern hybridization buffer: 6× SSC, 1% SDS, 10× Denhardt's solution.
3. A temperature-controlled rotary hybridization oven with bottles or hybridization trays with sealing bags.

R2.7–R2.10 describe materials for radioisotope-based method.

2.7 5′End-Labeling of the Probe with [γ-^{32}P]ATP

1. T4 Polynucleotide Kinase (PNK).
2. Gamma-^{32}P adenosine triphosphate ([γ-^{32}P] ATP) with high specific activity.
3. Sephadex G-25 spin columns.
4. Protective gear for radioactivity: Lab coat, gloves, and a Geiger counter.

2.8 Hybridization with Radiolabeled Probe

1. Prehybridization buffer (see Sect. 2.6) will be used.

2.9 Washing the Membrane

1. Medium stringency wash buffer: 2× SSC, 0.1% SDS
2. High stringency wash buffer: 0.1× SSC, 0.1% SDS

2.10 Signal Detection and Quantification

1. Phosphor screen
2. Exposure cassette
3. Phosphorimager
4. Stripping buffer: 0.1× SSC, 0.5% SDS

N2.7–N2.12 describe materials for nonradioactive method.

2.11 Preparation of DIG-Labeled Probe

1. DIG Oligonucleotide 3′-End Labeling Kit (#3353575910, Sigma-Aldrich). Alternatively, a biotin labeling kit (#89818, Thermo Scientific) can be used if biotin-streptavidin detection is preferred.
2. Sephadex G-25 spin columns (optional).

2.12 Hybridization with DIG-Labeled Probe

Use prehybridization buffer (see Sect. 2.6)

2.13 Washing the Membrane

1. Medium stringency wash buffer: 2× SSC, 0.1% SDS
2. High stringency wash buffer: 0.1× SSC, 0.1% SDS
3. Maleic acid buffer: 100 mM maleic acid, 150 mM NaCl, 0.3% Tween-20, pH 7.5

2.14 Blocking for Immunodetection

1. DIG Wash and Block Buffer Set (#11585762001, Roche)

2.15 Antibody Binding and Washing

1. Anti-Digoxigenin-AP, Fab fragments (#11093274910, Roche)
2. Membrane rinse buffer: 100 mM Tris-HCl, 100 mM NaCl, 50 mM $MgCl_2$, pH 9.5

2.16 Chemiluminescent Detection and Quantification

1. CDP-Star® (#11685627001, Roche)
2. Stripping buffer: 0.2 M NaOH, 0.1% SDS
3. Imaging system with cooled CCD camera

3 Methods

We first describe the setup through transfer (Steps 3.1–3.6), then provide separate procedures for radiolabeled probe detection (Steps R3.7–R3.10), and nonradioactive probe detection (Steps N3.7–N3.12).

3.1 RNA Preparation

1. Harvest cells.
2. Keep samples on ice and work RNase-free (use RNase inhibitors or DEPC-treated solutions).
3. Isolate total RNA using a suitable extraction method [14].
4. Verify RNA integrity by running an agarose gel.
5. Quantify RNA concentration.
6. Aliquot 5–20 μg total RNA per cleavage reaction to ensure sufficient substrate.
7. (Optional) If the target RNA is low-abundance, enrich it (e.g., poly(A) selection for mRNA).

8. Proceed immediately to cleavage reactions or store RNA appropriately.

3.2 In Vitro Cleavage Assay

1. Prepare nuclease-free tubes for the reactions.
2. Set the total reaction volume to 20 μL.
3. Add 5–20 μg of total RNA to each reaction.
4. Add 1× cleavage buffer.
5. Add the endonuclease of interest.
6. Use a dilution of the enzyme that results in cleavage of about 50% of the target RNA within the chosen time (this allows quantification of both cleaved and uncleaved fractions).
7. Prepare a negative control reaction without enzyme to distinguish specific cleavage products from background RNA breaks.
8. Incubate the reactions at the appropriate temperature for the required time.
9. Stop the reactions by adding an equal volume of stop/denaturing solution (2× RNA loading buffer containing formamide and EDTA).
10. If the enzyme requires divalent cations (e.g., Mg^{2+}), add 10 mM EDTA to chelate the ions and terminate the reaction.
11. Immediately chill the samples on ice.
12. Store the samples at −80°C or proceed directly to electrophoresis.

3.3 Denaturing Gel Electrophoresis

1. Thaw samples if they have been stored frozen.
2. Heat samples at 70°C for 3–5 min in formamide loading buffer to denature the RNA.
3. Quickly chill the samples on ice before loading onto the gel.
4. Load each reaction onto a denaturing gel prepared as described in the Materials section.
5. If analyzing endogenous cleavage events, load 10–20 μg of total RNA per lane.
6. Load an RNA size marker in a separate lane for reference.
7. For small fragments (<500 nt):
   - Use a urea-polyacrylamide gel.
   - Prerun the gel in 1× TBE for 10 min at the running voltage to warm it and remove urea contaminants.
   - Load samples and run at 200–300 V until the bromophenol blue dye approaches the bottom of the gel.

8. For larger fragments (>500 nt):
   - Use a denaturing agarose gel with formaldehyde.
   - Cast and set the gel in a fume hood.
   - Run at 100–150 V in 1× MOPS buffer until adequate separation is achieved (typically 1–2 h, depending on fragment size).
9. Verify migration after the run by visualizing RNA using ethidium bromide or SYBR Gold staining. Alternatively, add ethidium bromide directly to the samples before loading to visualize RNA migration during the run.

3.4 Membrane Transfer

1. Prepare the gel and membrane. After electrophoresis, disassemble the gel apparatus.
2. For polyacrylamide gels:
   - Carefully separate the glass plates and float the gel in 0.5× TBE to remove urea.
   - Semidry transfer option: Sandwich the gel and a prewetted nylon membrane between wet filter papers and transfer at constant current (0.5 mA/cm^2) for 1–2 h.
   - Tank transfer option: Transfer the gel in 0.5× TBE at 4°C using a tank electrotransfer system.
   - Ensure that the membrane is positioned on the cathode side of the gel so that RNA migrates onto it.
3. For agarose gels:
   - Briefly rinse the gel in water to remove residual formaldehyde.
   - Soak the gel in 10× SSC for 10 min.
   - Place the gel upside-down on a wick that bridges a dish containing 10× SSC.
   - Lay a prewetted nylon membrane (in 2× SSC) on top of the gel.
   - Cover with layers of wet blotting paper, then dry blotting paper, and place a weight on top.
   - Allow upward capillary transfer overnight to move RNA from the gel to the membrane.
4. Finalize the transfer:
   - After transfer, mark the lanes and orientation on the membrane with a pencil.
   - Gently rinse the membrane in 2× SSC to remove any remaining gel fragments.

3.5 RNA Fixation on Membrane

1. Keep the nylon membrane damp after transfer.
2. Lay the membrane face-up in a UV crosslinker.
3. Expose the membrane to 120 mJ/cm^2 UV to covalently attach nucleic acids to the nylon surface.
4. If the RNA fragments of interest are very small (<30 nt), consider using EDC (1-ethyl-3-(3-dimethylaminopropyl) carbodiimide) crosslinking instead (*see* **Note 7**).
5. Postcrosslink handling. After UV crosslinking, either:
   - Use the membrane immediately for hybridization, or
   - Air-dry the membrane for short-term storage.
6. Optional additional fixation. Some protocols recommend further securing nucleic acids by:
   - Baking the membrane for 30 min at 80°C, or
   - Air-drying the membrane overnight.

   However, UV crosslinking alone is usually sufficient.

3.6 Prehybridization

1. Put the nylon membrane into a hybridization bottle (for a rotary oven) or a flat tray (for sealed bags).
2. Cover the membrane with prewarmed ULTRAhyb-oligo or a homemade hybridization buffer.
3. Use approximately 0.1 mL of buffer per cm^2 of membrane (e.g., 10 mL for a standard 10 × 10 cm blot).
4. Place the bottle or tray in a hybridization oven with rotation or shaking to keep the solution evenly distributed.
5. Temperature settings.
   - For short oligonucleotide probes with ULTRAhyb-Oligo: incubate at ~42°C.
   - For longer probes in formamide-based buffers: incubate at 50°C–68°C, depending on probe Tm.
6. Incubate for at least 30 min to block nonspecific binding sites on the membrane.

3.7 5′ End-Labeling of the Probe with [γ-^{32}P]ATP

Follow steps R3.7–R3.10 for the ^{32}P-labeled probe method.

1. Place an RNase-free microtube on ice.
2. Pipette 30 pmol of the oligonucleotide probe (in water or TE buffer) into the tube.
3. Add 1 μL of 10× PNK reaction buffer.
4. Add 50 μCi of [γ-^{32}P]ATP.
5. Add 10 U T4 Polynucleotide Kinase (PNK).

6. Add nuclease-free water to bring the total volume to 10 μL.
7. Gently mix and quick-spin to collect the reaction at the bottom.
8. Incubate at 37°C for 60 min to allow 5′ end labeling.
9. Optional inactivation. Add 1 μL of 0.5 M EDTA to chelate Mg^{2+} and heat-inactivate at 65°C for 10 min (optional; can be omitted if proceeding directly).
10. Apply the reaction mixture to a pre-equilibrated Sephadex G-25 spin column.
11. Spin for 2 min at 700 g.
12. Retrieve the flow-through containing the purified radiolabeled probe.
13. Dilute the radiolabeled probe into hybridization buffer immediately before use.

3.8 Hybridization with Radiolabeled Probe

1. Pour off the prehybridization buffer from Step R3.6.
2. Add hybridization buffer containing the ^{32}P-labeled probe.
3. Rotate or gently agitate the bottle so that the probe solution distributes evenly across the membrane.
4. Hybridize the membrane overnight (8–16 h) at the chosen temperature.
   - Typical setting: ~42°C for short DNA oligonucleotide probes in ULTRAhyb-Oligo buffer.
   - Alternative ranges: Longer probes in formamide-based buffers may require higher temperatures (up to 50–68°C).
5. Sensitivity considerations.
   - For low-abundance targets, a long incubation (overnight or longer) maximizes probe binding.
   - For highly expressed targets, shorter incubations (a few hours) may be sufficient.

3.9 Washing the Membrane

1. Dispose of hybridization buffer. After hybridization, carefully remove the used buffer.
   - Treat it as radioactive waste, or
   - Optionally save it at −20°C for reuse (it can be reused multiple times).
2. First wash. Rinse the membrane in medium stringency wash buffer at room temperature for 5 min to remove unbound probe.
3. Second wash. Transfer the membrane to fresh medium stringency wash buffer and wash at 42°C for 15 min with gentle agitation.

4. Third wash. Repeat the second wash. If background remains, wash again in high stringency wash buffer at 42°C for 15 min.
5. Evaluate sufficiency.
   - Typically, two medium stringency washes are enough.
   - Use a Geiger counter after each wash: the membrane is ready once background radiation is low and localized to specific bands.
6. Remove wash buffer, and gently blot the membrane to remove excess liquid, keeping it moist but not dripping, in preparation for exposure.

3.10 Signal Detection and Quantification

1. Slightly air-dry the membrane, then place it against a sheet of plastic wrap. Seal it to prevent sticking to the film.
2. Mount the wrapped blot in an autoradiography cassette.
3. Place the membrane against a phosphorimaging screen for the appropriate exposure time.
4. Use a phosphorimager to obtain a digital image.
5. Measure the signal from bands corresponding to full-length RNA and cleaved fragments using image analysis software.
6. Ensure phosphorimager data are within the linear detection range.
7. Calculate cleavage fraction:

$$\text{Fraction cleaved} = \text{Intensity of cleavage fragment} / (\text{Intensity of cleavage fragment} + \text{full-length RNA})$$

8. Repeat for accuracy. Use multiple exposures if necessary and repeat experiments to ensure reproducibility.
9. To reprobe the blot, remove the ^{32}P probe by boiling the membrane in stripping buffer. Then incubate the membrane in fresh stripping buffer at 55°C for 1 h.
10. Check that no residual signal remains before rehybridizing with a new probe.

Follow steps N3.7–N3.12 for the DIG-labeled probe method.

3.11 Preparation of DIG-Labeled Probe

1. Setup the labeling reaction (3′ end labeling via terminal transferase):
   - Add 0.2 nmol of oligonucleotide.
   - Add 1 μL of 10× tailing buffer.
   - Add 1 μL of $CoCl_2$ solution (if provided).

- Add 2–5 µL of DIG-11-ddUTP or DIG-dUTP (typically 1 mM stock; follow kit instructions).
- Add 20 U of terminal deoxynucleotidyl transferase (TdT).
- Add nuclease-free water to adjust the total volume to 10 µL.

2. Allow the reaction to proceed at 37°C for 1 h.
3. Terminate the reaction.
   - Heat inactivate at 70°C for 10 min, or
   - Add EDTA as instructed in the kit protocol.
4. Optional purification. Precipitate or use a spin column to remove free DIG nucleotides. (This is often optional since hybridization buffer and washes remove excess label.)
5. Alternative method. For longer RNA probes, use in vitro transcription incorporating DIG-UTP [15].
6. Prepare working probe solution by diluting the DIG-labeled probe into hybridization buffer.
   - Typical usage: 10–50 ng of DIG-labeled probe per mL of buffer. Optimize empirically depending on target abundance.

3.12 Hybridization with DIG-Labeled Probe

1. Carefully discard the prehybridization buffer from Step 3.6.
2. Replace with fresh hybridization buffer containing the DIG-labeled probe.
3. Incubate the membrane at 42°C–50°C, depending on the probe composition and target RNA.
4. Place the bottle or tray in a hybridization oven with gentle rotation or agitation to ensure even probe distribution.
5. Hybridize the membrane overnight (8–16 h).
6. Sensitivity considerations.
   - For very low-abundance targets, probe enhancements (e.g., locked nucleic acids [LNAs]) may further improve sensitivity (*see* **Note 8**).
7. For most experiments, overnight hybridization balances sensitivity and practical workflow.

3.13 Washing the Membrane

1. Discard the used probe solution, or optionally save it at −20°C for reuse a few times.
2. First wash. Place the membrane in medium stringency wash buffer and wash at room temperature for 5 min with gentle shaking.
3. Second wash. Wash the membrane again in fresh medium stringency buffer at the hybridization temperature for 15 min.

4. Third wash. Repeat the medium stringency wash (same conditions as Step 3) one more time.
5. Optional high-stringency wash.
   - If the probe is short or background signal is high, wash the membrane in high stringency buffer (0.1× SSC, 0.1% SDS) at 50°C–55°C for 5–10 min.
6. Final rinse. Briefly rinse the membrane in maleic acid buffer to prepare for the blocking step.

3.14 Blocking Immunodetection

1. Lay the membrane flat in a clean tray.
2. Cover with DIG blocking buffer
   - The blocking reagent (casein or milk-derived protein) prevents nonspecific antibody binding.
   - Use ~0.1 mL of buffer per cm^2 of membrane surface.
3. Allow the membrane to incubate in the blocking buffer at 37°C for 30 min with gentle shaking.
   - Ensure the membrane is fully wetted and free-floating during this step.

3.15 Antibody Binding and Washing

1. Remove the blocking buffer from the tray.
2. Cover the membrane with the diluted anti-DIG-AP antibody solution prepared in fresh blocking buffer. Typical working dilutions: 1:5000 or 1:10,000 (optimize empirically).
3. Allow the membrane to incubate for 1 h at room temperature with gentle agitation to ensure even binding.
4. After incubation, discard the antibody solution or save it for reuse within the same day if processing multiple blots.
5. First wash. Place the membrane in DIG wash buffer and agitate gently for 10 min.
6. Second wash. Replace with fresh DIG wash buffer and wash again for 10 min.
7. Third wash. Repeat once more for 10 min.
8. Briefly rinse the membrane with rinse buffer
9. Do not allow the membrane to dry out at any point during these steps.

3.16 Chemiluminescent Detection and Quantification

1. Lay the washed membrane on a clean sheet of transparency film or plastic.
2. Pipette a few milliliters of chemiluminescent substrate for alkaline phosphatase (AP) onto the membrane surface.
3. Gently spread the substrate solution across the membrane, ensuring full coverage.
4. Leave the membrane at room temperature for 5 min to allow the AP enzyme to catalyze the reaction and generate light.

5. Drain off excess substrate carefully.
6. Seal the membrane in plastic wrap to prevent drying.
7. Place the membrane directly on a CCD-based chemiluminescence imaging system. Capture images digitally; this approach allows rapid signal quantification and avoids film saturation.
8. Calculate cleavage fraction:

$$\text{Fraction cleaved} = \text{Intensity of cleavage fragment} \\ /(\text{Intensity of cleavage fragment} + \text{full} \\ -\text{length RNA})$$

9. Linearity considerations.
   - Nonradioactive signals are generally linear within a certain range.
   - Avoid comparing bands if one is near the detection limit and another is very strong
10. Perform at least three independent experiments and report the mean ± SEM for cleavage percentages.
11. To reprobe the membrane, wash it in the stripping buffer for 15 min.
12. Rinse and equilibrate the membrane in 2× SSC.
13. Reblock the membrane with blocking buffer before applying a new probe.
14. Membranes can be reprobed multiple times with minimal loss of signal.

4 Notes

1. **RNase-Free Technique:** Throughout the experiment, take precautions to avoid RNase contamination. Use DEPC-treated water for solutions. Autoclave or filter-sterilize buffers when possible. Always wear gloves and use RNase-free pipette tips and tubes. It is good practice to wipe surfaces and gel apparatus with 70% ethanol or RNase decontamination solutions. Include RNase inhibitor in the cleavage reactions if compatible with the enzyme, especially if working with crude extracts.
2. **Optimizing Hybridization Stringency:** The hybridization and wash conditions should be optimized to maximize specific binding of the probe to the target fragment while minimizing non-specific background. Factors to adjust follows:

Temperature: Generally, hybridize 15°C–20°C below the melting temperature of the probe-target duplex. If a probe is not giving a signal, try lowering hybridization temperature a few degrees to allow more binding. If background is high, slightly increase temperature or reduce formamide percentage to increase stringency.

Salt and Detergent: Saline-Sodium Citrate (SSC) provides ionic strength for the DNA probe-RNA duplex formation. Higher salt such as 6× SSC during hybridization can increase sensitivity but also allow more off-target binding. SDS in the buffer help reduce nonspecific interactions. After hybridization, the stringency is typically increased by lowering salt from 2× to 0.1× SSC and raising temperature during washes. If you observe faint ghost bands or high background signals, additional low-salt washes might help.

Time: Longer hybridization up to 16 h generally increases sensitivity as it allows the probe to find and bind all target molecules. If time is limited, using a higher probe concentration can compensate to some extent for shorter hybridization, but at the risk of more background.

3. **Radioactive vs Nonradioactive Sensitivity:** Traditionally, ^{32}P-labeled probes were considered the most sensitive for Northern blots. However, with modern techniques, nonradioactive probes can achieve comparable sensitivity [11, 12]. For instance, chemiluminescent detection with DIG-labeled probes has been shown to detect as low as femtomole amounts of RNA, similar to or even better than ^{32}P in some cases. A study noted that a DIG-LNA probe could produce a detectable signal with just 0.05 fmol of target after only seconds of exposure, whereas a ^{32}P probe required overnight exposure for a similar signal [11]. In practice, if your target is very low abundance or if single-nucleotide resolution is needed, a radiolabeled probe on a high-resolution gel might still provide the clearest result. But for most applications, the difference in sensitivity is small, and nonradioactive methods offer significant convenience and safety advantages. It is wise to choose the method that best fits the laboratory's capabilities and safety considerations.

4. **Handling Radioactive Probes:** Follow radiation safety protocols when working with ^{32}P. Always use a Geiger counter to monitor your work area and dispose of waste in designated containers. Keep exposure short by working behind a shield, and store radioactive materials in lead or acrylic containers. After the experiment, decontaminate surfaces and conduct wipe tests if required by your institution. ^{32}P has a high energy β emission that can cause serious eye or skin damage. Thus, wear appropriate personal protective equipment. Also, given

the half-life of ^{32}P that is about 14 days, freshly label probes for each experiment to achieve maximal signal.

5. **Chemical Crosslinking for Small RNAs:** When detecting very small cleavage products less than 50 nt, the standard UV crosslinking can be inefficient because UV primarily crosslinks through thymidine residues in DNA or uridine in RNA to the membrane, and very short RNAs have few such residues or might dissociate. In such cases, you can use EDC (1-ethyl-3-(3-dimethylaminopropyl)carbodiimide) crosslinking, which covalently links the 3′-terminal riboses of RNA to the amine groups on the membrane [16]. The protocol involves soaking the membrane in 0.13 M EDC, 0.2 M 1-methylimidazole, pH 8.0, for 1 h at 50°C. This strongly immobilizes small RNAs and can improve their detection significantly. After EDC treatment, rinse the membrane, let it dry, and proceed to hybridization. This method was demonstrated to enhance detection of miRNAs and other tiny RNAs that UV cross-link poorly. One can combine this with UV crosslinking if analyzing a range of sizes.
6. **Safety and Waste Disposal:** Dispose of all hazardous waste according to local regulations. For radioactive experiments, this means collecting liquid scintillation waste, gel pieces, used membranes, and contaminated tips in proper waste drums. ^{32}P membranes can remain radioactive for a long time after imaging. Thus, either store them in a shielded container if you plan to reimage, or dispose of them as radioactive waste. Chemiluminescent substrates are generally nontoxic, but the buffers containing SDS and blocking reagents should be treated as lab chemical waste if they contain azide. Wear gloves when handling membranes after hybridization. If using UV transilluminators or crosslinkers, avoid direct UV exposure to skin and eyes.
7. **Alternative Probe Labeling (Biotin and Others):** While DIG is popular, biotin-labeled probes can be used similarly. Biotin can be incorporated at the end of an oligo in a similar way to the DIG labeling. The detection uses streptavidin or avidin conjugated to enzymes, such as horse radish peroxidase (HRP). One advantage is the strong affinity of streptavidin-biotin, but a disadvantage is that endogenous biotin in samples can sometimes cause background though usually not an issue on Northern since biotin is mostly in proteins/vitamins. The sensitivity of biotin-based Northern blot has been reported to be comparable to DIG in many cases [13]. If using biotin, blocking might need adjustment adding polyvinylpyrrolidone or sonicated salmon sperm DNA as blockers to reduce nonspecific binding of streptavidin. The procedure otherwise remains the same.

8. **Probe Design Considerations:** The choice of probe sequence and length can greatly affect sensitivity and specificity. Probes of 20–30 nucleotides are common for Northern blots; they strike a balance between having a high melting temperature and specificity. Avoid sequences with significant self-complementarity or repetitive elements. If possible, target a region near the cleavage site but entirely on one side of it so that the probe binds either only the upstream fragment or only the downstream fragment. If both cleavage products need detection, you can sequentially hybridize the blot with two different probes or run duplicate gels. For difficult targets such as GC-rich regions or very small RNAs, using locked nucleic acid (LNA)-modified probe significantly boosts affinity and stability of the probe-target duplex, thereby improving detection sensitivity [11]. LNA probes can be ordered or prepared by appropriate synthesis and are used in the same way in hybridization.

Acknowledgments

This work was supported by a National Research Foundation of Korea (NRF) grant funded by the Korean Ministry of Science (MSIT) (RS-2025-00556082 to J.H.), and the Brain Pool Program funded by the Korean Ministry of Science (MIST) and Information and Communication Technology (ICT) through the National Research Foundation of Korea (NRF) (RS-2024-00399517 to M.K.).

References

1. Gu K, Mok L, Chong MMW (2018) Regulating gene expression in animals through RNA endonucleolytic cleavage. Heliyon 4:e00908. https://doi.org/10.1016/j.heliyon.2018.e00908
2. Agrawal N et al (2003) RNA interference: biology, mechanism, and applications. Microbiol Mol Biol Rev 67:657–685. https://doi.org/10.1128/MMBR.67.4.657-685.2003
3. Aubert M, O'Donohue MF, Lebaron S, Gleizes PE (2018) Pre-ribosomal RNA processing in human cells: from mechanisms to congenital diseases. Biomolecules 8(4), 123. https://doi.org/10.3390/biom8040123
4. Phizicky EM, Hopper AK (2023) The life and times of a tRNA. RNA 29:898–957. https://doi.org/10.1261/rna.079620.123
5. Ha M, Kim VN (2014) Regulation of microRNA biogenesis. Nat Rev Mol Cell Biol 15:509–524. https://doi.org/10.1038/nrm3838
6. Carrard J, Lejeune F (2023) Nonsense-mediated mRNA decay, a simplified view of a complex mechanism. BMB Rep 56:625–632. https://doi.org/10.5483/BMBRep.2023-0190
7. Harigaya Y, Parker R (2010) No-go decay: a quality control mechanism for RNA in translation. Wiley Interdiscip Rev: RNA 1:132–141. https://doi.org/10.1002/wrna.17
8. Ahmad W, Gull B, Baby J, Mustafa F (2021) A comprehensive analysis of northern versus liquid hybridization assays for mRNAs, small RNAs, and miRNAs using a non-radiolabeled approach. Curr Issues Mol Biol 43:457–484. https://doi.org/10.3390/cimb43020036
9. Sloan KE et al (2013) Both endonucleolytic and exonucleolytic cleavage mediate ITS1 removal during human ribosomal RNA processing. J Cell Biol 200:577–588. https://doi.org/10.1083/jcb.201207131
10. Krumlauf R (1996) Northern blot analysis. Methods Mol Biol 58:113–128. https://doi.org/10.1385/0-89603-402-X:113
11. Kim SW et al (2010) A sensitive non-radioactive northern blot method to detect small

RNAs. Nucleic Acids Res 38:e98. https://doi.org/10.1093/nar/gkp1235

12. Madhry D, Roy R, Verma B (2024) Biotin-based northern blotting (BiNoB): a cost-efficient alternative for detection of small RNAs. Curr Protoc 4:e70065. https://doi.org/10.1002/cpz1.70065
13. Huang Q, Mao Z, Li S, Hu J, Zhu Y (2014) A non-radioactive method for small RNA detection by northern blotting. Rice (N Y) 7:26. https://doi.org/10.1186/s12284-014-0026-1
14. Rio DC, Ares M Jr, Hannon GJ, Nilsen TW (2010) Purification of RNA using TRIzol (TRI reagent). Cold Spring Harb Protoc 2010:pdb prot5439. https://doi.org/10.1101/pdb.prot5439
15. Sive HL, Grainger RM, Harland RM (2007) Synthesis and purification of digoxigenin-labeled RNA probes for in situ hybridization. CSH Protoc 2007:pdb prot4778. https://doi.org/10.1101/pdb.prot4778
16. Pall GS, Codony-Servat C, Byrne J, Ritchie L, Hamilton A (2007) Carbodiimide-mediated cross-linking of RNA to nylon membranes improves the detection of siRNA, miRNA and piRNA by northern blot. Nucleic Acids Res 35:e60. https://doi.org/10.1093/nar/gkm112

Check for updates

Chapter 4

Molecular Analysis of miRNA Stability and Decay Dynamics by Northern Blot

Kyeong-Ho Eun, Ho-Chang Jeong and Jaeil Han

Abstract

MicroRNAs (miRNAs) are critical post-transcriptional regulators that function through association with Argonaute proteins to form the RNA-induced silencing complex (RISC). Although miRNAs were considered remarkably stable molecules, their cellular abundance is subject to active regulation through coordinated degradation pathways. Precise quantification of miRNA stability remains technically challenging, primarily owing to the small size of these molecules, their elevated GC content, and their relatively low abundance compared to other cellular RNA species. To overcome these technical limitations, this chapter provides a comprehensive experimental strategy for monitoring miRNA stability using optimized northern blotting techniques. This approach integrates transcriptional shutoff through actinomycin D treatment with advanced detection chemistries—specifically, EDC-mediated crosslinking and near-infrared fluorescent probe hybridization—to markedly enhance signal resolution and detection sensitivity, thereby facilitating precise determination of decay kinetics and reliable calculation of miRNA half-lives. Collectively, this method offers high specificity, enhanced sensitivity, and quantitative accuracy to investigate miRNA degradation dynamics at single-species resolution.

Keywords miRNA decay, Actinomycin D, EDC crosslinking, Near-infrared fluorescent probe hybridization, Northern blotting

1 Introduction

MicroRNAs (miRNAs), a class of small non-coding RNAs typically 19–25 nucleotides in length, function as fundamental post-transcriptional regulators of gene expression in eukaryotic systems [1, 2]. These molecules guide Argonaute (AGO) proteins, which serve as central catalytic components of the RNA-induced silencing complex (RISC), to target messenger RNAs [3–8]. Upon recognition and binding to partially complementary target sequences, miRNAs induce target messenger RNA degradation and translational repression [9–11]. Therefore, it is crucial to understand the molecular basis of miRNA turnover, whose abundance is not solely determined by

Je-Hyun Yoon (ed.), *RNA Labeling*, Methods in Molecular Biology, vol. 3047,
https://doi.org/10.1007/978-1-0716-5352-4_4,

synthesis rates but is profoundly shaped by active, regulated degradation pathways [12–15].

Investigation of miRNA decay has been hampered by substantial technical obstacles, particularly due to the small molecular size of miRNAs, their elevated GC content, relatively low cellular abundance compared to other RNA species, and remarkable stability within the AGO complex [2, 16–20]. To study mature miRNA stability, transcription-blocking agents such as actinomycin D were widely applied to prevent cells from synthesizing fresh RNA transcripts and pri-miRNA precursors [21–23]. Combined actinomycin D treatment with quantitative reverse transcription PCR enables the monitoring of individual miRNA species decay, particularly when specific probes are designed to discriminate among primary, precursor, and mature miRNA forms [13, 24]. Alternatively, metabolic RNA labeling with modified nucleosides (4-thiouridine 4sU or 5-ethynyl uridine 5-EU) has been developed to complement transcriptional shut-off assays [25–27]. These nucleotide analogs integrate into nascent RNA molecules, allowing the tracking of RNA turnover through bromouridine immunoprecipitation coupled with sequencing (BRIC-seq) [25, 26]. However, these approaches require technical expertise and may lack sufficient sensitivity for low-abundance small RNAs or their precursor forms, although they are powerful for transcriptome-wide RNA dynamics analysis [12, 13, 25, 27–30]. Reporter-based assays, linking miRNA target sites to reporter gene expression, have also been employed to indirectly assess miRNA stability by monitoring changes in their reporter activity reflecting miRNA-mediated regulation [11, 31–34]. Nonetheless, a significant hurdle in these conventional approaches remains the difficulty in discriminating mature miRNAs from precursor forms (pre-miRNAs and pri-miRNAs) or degradation intermediates, especially at low abundance [2, 17–19, 35, 36].

To overcome these technical limitations, this book chapter provides optimized and highly sensitive approaches for quantifying miRNA decay dynamics [13, 37, 38]. Specifically, we introduce refined northern blotting protocols that, when combined with transcriptional shutoff using actinomycin D, effectively resolve and quantify both mature miRNAs and their precursor forms [21, 37, 38]. The protocol describes the applications of enhanced detection strategies, including electrochemiluminescence crosslinking for signal amplification and resolution, and near-infrared fluorescent probe hybridization for highly sensitive and quantitative detection [37, 38]. Collectively, this integrated approach is enabling precise quantification of miRNA degradation rates to understand the impact of miRNA stability on diverse biological processes through the regulation of gene expression [13, 28, 37]. Detailed experimental procedures, reagents, and critical technical considerations are provided to enable efficient and reproducible performance of the protocol (Fig. 1).

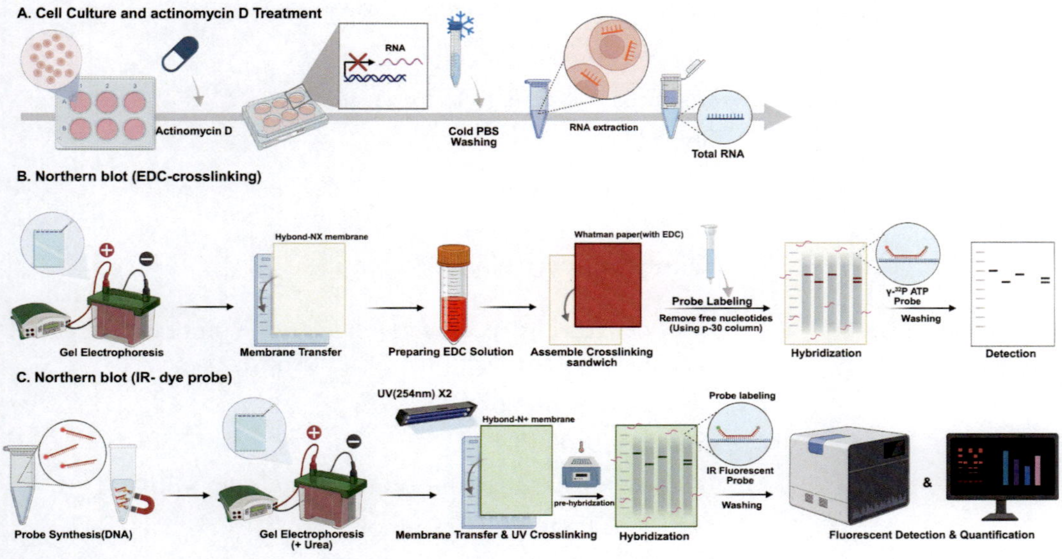

Fig. 1 Schematic representation of miRNA decay analysis using northern blot. (**A**) Cells were treated under the indicated conditions, and total RNA was isolated at defined time points to monitor miRNA stability. (**B**) Workflow of northern blot analysis using EDC-mediated crosslinking. Total RNA was resolved on denaturing polyacrylamide gels, transferred onto Hybond-NX membranes, prepared EDC solution, and hybridized with a γ-^{32}P-labeled DNA probe specific to the target miRNA. Signals were visualized by phosphorimaging. (**C**) Workflow of infrared dye (IR)-based Northern blot analysis. DNA probes were first synthesized and conjugated with IR Dye DECO (680RD or 800CW) via click chemistry. After pre-hybridization, total RNA samples were separated by electrophoresis and UV crosslinked onto Hybond-N$^+$ membranes. The IR-labeled probes were then hybridized to target miRNAs, and fluorescent signals were detected and quantified using a Li-Cor Odyssey imaging system

2 Materials

2.1 Cell Culture and Actinomycin D Treatment

1. TTP-deficient and wild-type mouse embryonic fibroblasts (MEFs).
2. Mouse embryonic stem cells (mESCs).
3. ES cell-grade fetal bovine serum (FBS) (Sigma, #F9423).
4. Knock-Out DMEM (Gibco, #10829).
5. L-Glutamine, Penicillin-Streptomycin.
6. Leukemia inhibitory factor (LIF).
7. b-Mercaptoethanol, DMEM.
8. Actinomycin D.
9. Dimethyl sulfoxide (DMSO).
10. Phosphate-buffered saline (PBS).
11. 60-mm cell culture dishes and 6-well culture plates.

2.2 RNA Extraction

1. TRI reagent (MRC, #TR 118).
2. Chloroform, Isopropanol.
3. 75% Ethanol (in RNase-free water).
4. Glycogen.
5. RNase-free water.
6. DNase I.
7. NanoDrop spectrophotometer.

2.3 Northern Blot Analysis (EDC-Crosslink)

1. 15% denaturing polyacrylamide gel solution (UreaGel-SequaGel System; National Diagnostics).
2. 0.5× TBE running buffer.
3. Ethidium bromide.
4. Oligo-hyb hybridization solution (Ambion, #8047-1).
5. γ-[^{32}P] ATP (>7000 Ci/mmol).
6. T4 polynucleotide kinase (PNK).
7. P-30 spin columns.
8. Amersham Hybond-NX membrane (GE Healthcare, #RPN68D).
9. Whatman 3MM chromatography paper.
10. EDC (1-ethyl-3-(3-dimethylaminopropyl) carbodiimide).
11. 10% APS, TEMED.
12. Wash buffer: 100 mL of 20× SSC, 100 mL of 10% SDS, and 800 mL of H_2O.
13. 2× RNA loading dye (Thermo Fisher, #R0641).
14. Stripping buffer: 2 mL of 10% SDS and 498 mL of H_2O.
15. 1× methylimidazole buffer: 408 μL of 12.5 M 1-methylimidazole to 30 mL water, adjust the pH to 8.0 with 1 M HCl.
16. DNA oligonucleotide (~20 mer, 20 pmoles).
17. 0.22 μm filter.
18. Hybridization tube (Life Science, #7995-150).
19. Blotting filter paper.
20. Gel electrophoresis apparatus (0.8 mm spacers, 19.5 × 16 cm plates).
21. Hybridization oven (with rotation).
22. Bio-Rad Trans-Blot SD semidry transfer unit (Bio-Rad, #1703848).
23. UV transilluminator.
24. Phosphorimager screen and cassette.
25. Gel documentation system.
26. Geiger counter.

2.4 Northern Blot Analysis (IR Dye-Probe)

1. 1.5 mL Seal-Rite microcentrifuge tubes (USA Scientific, #1615-5599).
2. Hybond-N$^+$ membrane, 30 cm × 3 m roll (GE Healthcare, #RPN303B).
3. Glycogen.
4. Elution buffer: 300 mM NaOAc and 25 mM Tris-HCl, pH 8.0.
5. Urea.
6. Oligo-hyb hybridization solution (Ambion, #8047-1).
7. Phenol, Isoamyl alcohol.
8. AMPure XP magnetic beads (Beckman Coulter, # A63881).
9. 5-Azidomethyl-dUTP (Jena Biosciences, #CLK-084S).
10. Terminal transferase (with $CoCl_2$) (New England Biolabs, #M0315).
11. 1× TdT buffer, 0.25 mM $CoCl_2$, 100 pmol DNA oligo (without azide).
12. IRDye 680RD DBCO and IRDye 800CW DBCO (Li-Cor, #929-50005 and #929-55000).
13. SYBR Green I nucleic acid gel stain.
14. 10% UltraPure™ SDS solution.
15. 1× PBS (pH 7.4): 137 mM NaCl, 10 mM Na2HPO4, and 2.7 mM KCl.
16. let-7a and U6 azide-modified DNA probes (ordered from IDT, HPLC-purified).
17. 8 M urea 15% polyacrylamide gel: 5 g of Urea, 2 ml of 5X TBE, 625 μl of ddH2O, 3.75 ml of 40% Acrylamide, 52 μl of 10% APS, and 10 μl of TEMED.
18. 2X formamide loading dye: 80% Formamide, 0.1% Bromophenol blue, 0.1% Xylene cyanol, 20 mM EDTA, and 20 mM Tris–HCl, pH 8.0.
19. DNA oligonucleotide (without azide).
20. UltraPure™ DNase/RNase-free distilled water.
21. ThermoMixer C, Magnetic rack.
22. UV crosslinker (254 nm, #Spectroline XL1000).
23. Odyssey CLX imaging system (Li-Cor).
24. Mini-PROTEAN Tetra gel unit.
25. Owl HEP series semidry blotter.
26. ImageQuant 300 imager (GE Healthcare).
27. Vortex, Rotating mixer.

3 Methods

3.1 Cell Culture and Actinomycin D Treatment

1. Seed mouse embryonic stem cells (mESCs) at a density of 5 × 10^5 cells per 60-mm gelatin-coated culture dish and incubate in Knock-Out DMEM supplemented with 15% FBS, 2 mM L-Glutamine, 0.1 mM β-Mercaptoethanol, and 1000 U/ml Leukemia inhibitory factor (LIF).
2. Plate mouse embryonic fibroblasts (MEFs) in 6-well culture plates at 2.5 × 10^5 cells per well and maintain in DMEM supplemented with 10% FBS and 0.1% Penicillin-Streptomycin.
3. Before actinomycin D (ActD) treatment, refresh culture media every two days, ensuring cells reach 70%–80% confluence.
4. Introduce ActD into the medium at a final concentration of 10 μg/mL to inhibit transcription (*see* **Note 1**).
5. Treat control cells with an equivalent volume of DMSO.
6. Harvest Cells at various time points following ActD treatment, including 0, 1, 2, 4, 6, 8, 10, 12, 18, and 24 h.
7. At each designated time point, aspirate the culture medium, and wash cells twice with ice-cold PBS before collection for subsequent RNA extraction (*see* **Note 2**).

3.2 Total RNA Extraction and Quality Control

1. Extract total RNA from the harvested cells using the TRI reagent.
2. Briefly, homogenize cell pellets in 1 mL of the TRI reagent and incubate at room temperature for 5 min.
3. Add 200 μL of chloroform, mix thoroughly, and centrifuge the samples at 12,000 × *g* for 15 min at 4°C to separate phases.
4. Transfer the aqueous phase containing RNA, add 500 μL of isopropanol to precipitate RNA, and incubate samples at −20 °C for at least 1 h (*see* **Note 3**) followed by centrifugation at 12,000 × *g* for 10 min at 4°C. A yellowish pellet should be visible.
5. Remove isopropanol carefully, and wash precipitated RNA with 1 mL of 75% ethanol by spinning at 12,000 × *g* for 5 min at 4 °C.
6. Remove ethanol, and centrifuge sample at 12,000 × *g* for 1 min followed by complete removal of residual ethanol.
7. Resuspend RNA pellet in 50–100 μL of RNase-free water.
8. Remove residual genomic DNA by DNase I treatment (*see* **Note 4**).
9. Determine RNA concentration spectrophotometrically using a NanoDrop instrument, and assess RNA purity (A260/A280 ratio) and integrity (agarose gel electrophoresis).

3.3 Northern Blot Analysis (EDC-Crosslink)

3.3.1 Gel Electrophoresis

1. Prepare one 15% denaturing polyacrylamide gel (18 mL concentrate, 9 ml diluent, 3 mL buffer, 100 μL APS (from 10% stock), 30 μL TEMED) (*see* **Note 5**).
2. Pre-run the gel at 5 W for 20 min (*see* **Note 6**).
3. Prepare RNA samples. Mix 5–10 μg of total RNA (≤15 μL) with an equal volume of 2× loading dye.
4. Heat at 80°C for 5–10 min, then spin briefly.
5. Load samples with a beveled pipette tip onto the pre-warmed gel.
6. Run initially at 2 W for 20 min to allow uniform entry and then increase to 12 W for approximately 2 h until the bromophenol blue dye exits the gel (Xylene cyanol should remain mid-gel) (for post-run check, *see* **Note 7**).

3.3.2 RNA Transfer to Membrane

1. Cut an Amersham Hybond-NX membrane and five sheets of Whatman 3MM paper slightly larger than the gel (Gel expands after staining). Typically, 15 cm × 18.5 cm fits one standard gel.
2. Label the membrane with pencil (date, experiment name) and cut a corner for orientation.
3. Pre-soak the membrane and filter papers in 0.5× TBE buffer.
4. Lay the membrane on the gel (still on plastic wrap) and align the edges. Trim any excess gel.
5. Assemble the sandwich. Gel–membrane–two filter papers (from bottom to top), then invert and place in the semidry transfer unit.
6. Add two additional filter papers on top, smooth out bubbles using a pipette roller, and ensure full moisture with 0.5× TBE (*see* **Note 8**).
7. Transfer at constant current (4 mA/cm^2 of membrane; ~900 mA total for one full gel) for 35 min. The voltage typically remains near 20 V. Do not exceed 40 min to prevent membrane overheating and loss of signal.

3.3.3 EDC Crosslinking

The EDC crosslinking method described here is adapted from the revised protocol originally developed in the Bartel Lab for small RNA detection [2].

1. Prepare 1× methylimidazole buffer, bring volume to 40 mL, and filter it to sterilize.
2. For each membrane, dissolve 0.466 g of EDC in 15 mL of 1× methylimidazole buffer immediately before use (*see* **Note 9**).

3. After the transfer was completed, the gel sandwich was disassembled, and the gel was soaked in ethidium bromide staining solution for approximately 5 min to verify successful RNA transfer. A faint smear of residual material near the wells was occasionally observed; however, distinct rRNA and tRNA bands were no longer visible, confirming efficient transfer. The stained gel was subsequently imaged for documentation.
4. For RNA crosslinking, a sheet of Whatman paper soaked in EDC solution was placed on plastic wrap, and the membrane was positioned RNA side down. The assembly was sealed tightly and incubated at 60°C for 1 h (15 min–2 h range).
5. Rinse the membrane several times in distilled water.
6. The membrane was immediately subjected to pre-hybridization.

3.3.4 Pre-Hybridization, Probe Labeling, and Hybridization

1. Warm Oligo-hyb solution to 60°C until fully thawed.
2. Roll the membrane with the RNA side facing inward and place it in a hybridization tube. Ensure no folds or creases.
3. Add ~15 mL of pre-warmed Oligo-hyb buffer per blot.
4. Rotate at 42°C for at least 30 min (Overnight pre-hybridization is also possible).
5. Prepare a 10 uL of T4 PNK labeling reaction as follows: 1 μL of 10 μM DNA oligo (~20 nt; 20 pmol), 1 μL of 10X PNK buffer, 1 μL of γ-[^{32}P] ATP (>7000 Ci/mmol), 1 μL of T4 polynucleotide kinase, and 6 μL of nuclease-free water.
6. Incubate the mixture at 37 °C for 1 h.
7. Add 50 μL of water and purify the labeled probe with a P-30 spin column following the manufacturer's instructions. Typically, ~30%–35% labeling efficiency is achieved.
8. For hybridization, add 0.5 mL of Oligo-hyb buffer to the labeled probe, heat at 85°C for 5 min, then introduce directly into the pre-hybridization solution (Avoid direct contact with the membrane).
9. Hybridize overnight (no more than 16 h) at 42°C to prevent high background (*see* **Note 10**).

3.3.5 Washing and Exposure

1. Prepare 1 L of wash buffer.
2. Collect hybridization buffer and store at −20 °C (Probe can be reused) (*see* **Note 11**).
3. Wash membrane twice with 30 mL of wash buffer for 5 min each, followed by one 30 min wash at room temperature.
4. Wrap the membrane in plastic wrap supported by Whatman paper, label the back, and check the signal using a Geiger counter.

5. Expose the blot to a phosphorimager screen for 30 min to 1 h initially, repeat with longer exposure for overnight to enhance sensitivity.

3.3.6 Stripping the Membrane

1. Prepare 500 mL of stripping buffer.
2. Heat to boiling in a microwave (Avoid excessive boiling).
3. Pour the hot solution over the blot (RNA side up) in a glass dish and agitate gently for 20 min.
4. Repeat twice more.
5. Wrap and expose the stripped blot briefly to a phosphorimager screen to confirm probe removal.

3.4 Northern Blot Analysis (IR Dye Probe)

The synthesis of IR dye-labeled oligonucleotide probes was carried out following the procedure [37], which describes conjugation of azide-modified oligonucleotides with DBCO-linked IRDyes (Li-Cor) for nonradioactive northern blot hybridization.

3.4.1 Preparation of IR-Labeled DNA Probe

1. Azide-functionalized DNA oligonucleotides were dissolved in double-distilled water (ddH_2O) to a final concentration of 100 μM.
2. Combine 2.5 nmol of DNA oligo with 50 nmol of IRDye DBCO (680RD or 800CW) in 1× PBS to a total volume of 50 μL (*see* **Note 12**).
3. Incubate the mixture at room temperature for approximately 6 h without agitation or shaking. Longer incubation (up to overnight) does not reduce yield.
4. Add 2 volumes of AMPure XP beads and 5.4 volumes of isopropanol, mix thoroughly by pipetting, and incubate for 7.5 min. Repeat mixing and incubation once more.
5. Collect beads on a magnetic rack, remove the supernatant, and wash twice with 85% ethanol.
6. Elute DNA in 100 μL of ddH_2O and transfer to a new tube.
7. Measure 1 μL using a NanoDrop spectrophotometer. Typical recovery efficiency is 70%–90%.

3.4.2 Alternative DNA Probe Labeling via Terminal Transferase

1. Assemble a 30 μL reaction containing 1× TdT buffer, 0.25 mM $CoCl_2$, 100 pmol DNA oligo (without azide), 300 pmol 5-Azidomethyl-dUTP, 20 U terminal transferase, and water to volume (*see* **Note 13**).
2. Incubate at 37°C for 1 h without shaking.
3. Purify the reaction using AMPure XP beads as in Sect. 3.4.1 and elute in 20 μL water.
4. Mix 19 μL of purified probe with 10 nmol IRDye DBCO in 1× PBS and incubate at room temperature for 6 h in the dark.

5. Purify again with AMPure XP beads as in Sect. 3.4.1 and elute in 15 μL water.
6. The entire IR-labeled oligonucleotide eluate was analyzed on an 8 M urea, 15% polyacrylamide gel, alongside 10 pmol of unlabeled oligonucleotide. The unlabeled oligo served as a reference to estimate the number of nucleotides added during the TdT reaction.
7. Visualize IRDye signal using the Li-Cor Odyssey scanner and stain the gel with SYBR Green I to identify fully labeled probes (*see* **Note 14**).
8. Excise the desired band and elute overnight in 600 μL of mixed buffer (mixed 1:1 with phenol/chloroform/isoamyl alcohol and elution buffer) on a rotating mixer.
9. Transfer the liquid to a new 1.5 mL tube, vortex for 30 s, and centrifuge at 21,000 × *g*, 4°C for 5 min.
10. Transfer the resulting aqueous phase to a fresh 1.5 mL tube.
11. Add 300 μL of chloroform, vortex for 30 s, and centrifuge at 21,000 × *g*, 4°C for 5 min.
12. Transfer the aqueous phase to a new 1.5 mL tube.
13. Add 15 μg of glycogen to aid co-precipitation and visualization of the pellet.
14. Add 1 mL cold 100% ethanol and incubate at −80 °C for 15 min.
15. Centrifuge at 21,000 × *g*, 4 °C for 30 min.
16. Remove the supernatant and air-dry the pellet for ~3 min or until dry.
17. Resuspend probe in 22 μL of ddH$_2$O and quantify it using a NanoDrop spectrophotometer.

3.4.3 Urea-Page, Transfer, and IR-Northern Hybridization

1. Mix 15–20 μg of RNA with an equal volume of 2× formamide loading dye, heat to 95°C for 3 min, then chill on ice.
2. Resolve RNAs on a 15% polyacrylamide gel containing 8 M urea.
3. Transfer the RNA to a Hybond-N$^+$ membrane using a semidry transfer unit at 300 mA for 1 h.
4. Crosslink the membrane twice with 254 nm UV at 120 mJ/cm^2.
5. Pre-hybridize the membrane in 10 mL of Oligo-hyb buffer at 30 °C for 30 min in a hybridization oven.
6. Hybridize overnight at 30°C with 10 pmol IR-labeled probe in 10 mL of Oligo-hyb solution.

7. Wash the membrane in 2× SSC with 0.1% SDS for 10 min at room temperature with gentle shaking, then in 1× SSC with 0.1% SDS for 10 min (*see* **Notes 15 and 16**).
8. Air-dry the membrane on paper towels and image using a Li-Cor Odyssey scanner (700 nm and/or 800 nm channels).
9. To strip the probe, immerse the membrane in boiled 0.1X SSC with 1% SDS and 40 mM Tris–HCl (pH 8.0) for 10 min at room temperature with gentle shaking. Repeat once.
10. Dry the membrane and store it in a sheet protector for future reuse (*see* **Note 17**).

4 Note

1. This concentration of actinomycin D is commonly used to effectively inhibit transcription in mammalian cells without immediately inducing significant cytotoxicity, allowing for the observation of RNA decay kinetics over several hours [23, 39].
2. Immediate aspiration and washing with ice-cold PBS help to halt cellular processes and minimize RNA degradation during sample collection.
3. Incubation at −80 °C for at least 1 h, or preferably overnight, enhances RNA precipitation efficiency, particularly for low-abundance RNAs.
4. DNase I treatment is optional to eliminate contaminating genomic DNA, thereby minimizing nonspecific probe binding in the subsequent northern blot.
5. After the gel has set, carefully remove the combs to ensure that the wells remain flat and undamaged. Mount the gel onto the electrophoresis rig and gently pour 0.5× TBE running buffer at an angle to avoid disturbing the wells. Make sure to remove any trapped air bubbles at the bottom of the gel. For optimal heat distribution and consistent band migration across lanes, an aluminum plate backing may be used during electrophoresis.
6. Before loading samples, gently rinse the wells with running buffer to remove any residual urea that may have accumulated during pre-run or gel preparation. This helps prevent sample distortion and ensures even loading across all wells.
7. Post-run check

   7.1. Transfer the gel onto plastic wrap, remove wells, and mark the first lane.

7.2. Stain in 4 μg/mL ethidium bromide (36 μL of stock in 90 mL 0.5× TBE) for 5 min.

7.3. Visualize tRNA and 5S rRNA bands (~78 nt and 120 nt) as loading and quality control.

7.4. Record an image before transfer for documentation.

8. Avoid air bubbles between gel and membrane during transfer, as these cause local loss of signal.
9. Use fresh EDC solution for each crosslinking step. Old solution significantly reduces coupling efficiency. RNA is covalently coupled to nylon membranes via carbodiimide chemistry using EDC and methylimidazole buffer.
10. The hybridization temperature was adjusted based on the GC content of the probe.
11. Hybridization buffer and probes can be reused multiple times if properly stored.
12. Keep all IRDye reagents protected from light.
13. The reaction can be scaled up for large-scale probe preparation. However, initial small-scale TdT reactions with varying probe-to-5-azidomethyl-dUTP ratios are recommended to assess labeling efficiency.
14. Multiple probes labeled with different IRDyes (680RD/800CW) can be used simultaneously for multi-target detection.
15. The washing temperature may be adjusted according to the melting temperature (Tm) of each probe–target pair to optimize hybridization stringency.
16. To increase washing stringency, a washing buffer can be used with reduced SSC concentration and elevated SDS concentration.
17. Dried membranes are stable for long-term storage and can be re-probed after stripping.

Acknowledgments

This work was supported by the National Research Foundation of Korea Grant funded by the Ministry of Science and ICT and the Ministry of Health and Welfare (Grant No. RS-2024-00341655 and RS-2024-00432867 to H-C. J, and RS-2025-00556082 to J.H).

References

1. Bartel DP (2004) MicroRNAs: genomics, biogenesis, mechanism, and function. Cell 116:281–297. https://doi.org/10.1016/s0092-8674(04)00045-5
2. Fang W, Bartel DP (2015) The menu of features that define primary microRNAs and enable de novo design of microRNA genes. Mol Cell 60:131–145. https://doi.org/10.1016/j.molcel.2015.08.015
3. Ender C, Meister G (2010) Argonaute proteins at a glance. J Cell Sci 123:1819–1823. https://doi.org/10.1242/jcs.055210
4. Liu J, Carmell MA, Rivas FV et al (2004) Argonaute2 is the catalytic engine of mammalian RNAi. Science 305:1437–1441. https://doi.org/10.1126/science.1102513
5. Schirle NT, Sheu-Gruttadauria J, MacRae IJ (2014) Structural basis for microRNA targeting. Science 346:608–613. https://doi.org/10.1126/science.1258040
6. Ergin K, Çetinkaya R (2022) Regulation of MicroRNAs. Methods Mol Biol 2257:1-32. https://doi.org/10.1007/978-1-0716-1170-8_1
7. Frédérick PM, Simard MJ (2022) Regulation and different functions of the animal microRNA-induced silencing complex. Wiley Interdiscip Rev: RNA 13:e1701. https://doi.org/10.1002/wrna.1701
8. Nakanishi K (2016) Anatomy of RISC: how do small RNAs and chaperones activate argonaute proteins?. Wiley Interdiscip Rev: RNA 7:637–660. https://doi.org/10.1002/wrna.1356
9. Bazzini AA, Lee MT, Giraldez AJ (2012) Ribosome profiling shows that miR-430 reduces translation before causing mRNA decay in zebrafish. Science 336:233–237. https://doi.org/10.1126/science.1215704
10. Djuranovic S, Nahvi A, Green R (2012) miRNA-mediated gene silencing by translational repression followed by mRNA deadenylation and decay. Science 336:237–240. https://doi.org/10.1126/science.1215691
11. Filipowicz W, Bhattacharyya SN, Sonenberg N (2008) Mechanisms of post-transcriptional regulation by microRNAs: are the answers in sight?. Nat Rev Genet 9:102–114. https://doi.org/10.1038/nrg2290
12. Großhans H, Chatterjee S (2010) MicroRNAses and the regulated degradation of mature animal miRNAs. Adv Exp Med Biol 700:140–155. https://doi.org/10.1007/978-1-4419-7823-3_12
13. Guo Y, Liu J, Elfenbein SJ et al (2015) Characterization of the mammalian miRNA turnover landscape. Nucleic Acids Res 43:2326–2341. https://doi.org/10.1093/nar/gkv057
14. Rüegger S, Großhans H (2012) MicroRNA turnover: when, how, and why. Trends Biochem Sci 37:436–446. https://doi.org/10.1016/j.tibs.2012.07.002
15. Xie M, Zhang S, Yu B (2015) microRNA biogenesis, degradation and activity in plants. Cell Mol Life Sci 72:87–99. https://doi.org/10.1007/s00018-014-1728-7
16. Ballarino M, Pagano F, Girardi E et al (2009) Coupled RNA processing and transcription of intergenic primary microRNAs. Mol Cell Biol 29:5632–5638. https://doi.org/10.1128/MCB.00664-09
17. Chugh P, Dittmer DP (2012) Potential pitfalls in microRNA profiling. Wiley Interdiscip Rev: RNA 3:601–616. https://doi.org/10.1002/wrna.1120
18. Dong H, Lei J, Ding L et al (2013) MicroRNA: function, detection, and bioanalysis. Chem Rev 113:6207–6233. https://doi.org/10.1021/cr300362f
19. Pritchard CC, Cheng HH, Tewari M (2012) MicroRNA profiling: approaches and considerations. Nat Rev Genet 13:358–369. https://doi.org/10.1038/nrg3198
20. Winter J, Diederichs S (2011) Argonaute proteins regulate microRNA stability: Increased microRNA abundance by Argonaute proteins is due to microRNA stabilization. RNA Biol 8:1149–1157. https://doi.org/10.4161/rna.8.6.17665
21. Levy HB (1963) Effect of actinomycin D on heLa cell nuclear RNA metabolism. Proc Soc Exp Biol Med 113:886–889. https://doi.org/10.3181/00379727-113-28521
22. Perry RP, Kelley DE (1970) Inhibition of RNA synthesis by actinomycin D: characteristic dose-response of different RNA species. J Cell Physiol 76:127–139. https://doi.org/10.1002/jcp.1040760202
23. Ratnadiwakara M, Änkö M-L (2018) mRNA stability assay using transcription inhibition by actinomycin D in mouse pluripotent stem cells. Bio-protoc 8:e3072–e3072. https://doi.org/10.21769/BioProtoc.3072
24. Schmittgen TD, Lee EJ, Jiang J et al (2008) Real-time PCR quantification of precursor and mature microRNA. Methods 44:31–38. https://doi.org/10.1016/j.ymeth.2007.09.006
25. Duffy EE, Rutenberg-Schoenberg M, Stark CD et al (2015) Tracking distinct RNA populations using efficient and reversible covalent chemistry. Mol Cell 59:858–866. https://doi.org/10.1016/j.molcel.2015.07.023
26. Imamachi N, Tani H, Mizutani R et al. (2014) BRIC-seq: a genome-wide approach

for determining RNA stability in mammalian cells. Methods 67:55–63. https://doi.org/10.1016/j.ymeth.2013.07.014

27. Szabo EX, Reichert P, Lehniger M-K et al (2020) Metabolic labeling of RNAs uncovers hidden features and dynamics of the Arabidopsis transcriptome. Plant Cell 32:871–887. https://doi.org/10.1105/tpc.19.00214
28. Marzi MJ, Ghini F, Cerruti B et al (2016) Degradation dynamics of microRNAs revealed by a novel pulse-chase approach. Genome Res 26:554–565. https://doi.org/10.1101/gr.198788.115
29. Tani H, Mizutani R, Salam KA et al (2012) Genome-wide determination of RNA stability reveals hundreds of short-lived noncoding transcripts in mammals. Genome Res 22:947–956. https://doi.org/10.1101/gr.130559.111
30. Yamada T, Akimitsu N (2019) Contributions of regulated transcription and mRNA decay to the dynamics of gene expression. Wiley Interdiscip Rev: RNA 10:e1508. https://doi.org/10.1002/wrna.1508
31. Clément T, Salone V, Rederstorff M (2015) Dual luciferase gene reporter assays to study miRNA function. Method Mol Biol 1296:187–198. https://doi.org/10.1007/978-1-4939-2547-6_17
32. Hendrickson DG, Hogan DJ, McCullough HL et al (2009) Concordant regulation of translation and mRNA abundance for hundreds of targets of a human microRNA. PLoS Biol 7:e1000238. https://doi.org/10.1371/journal.pbio.1000238
33. Larsson E, Sander C, Marks D (2010) mRNA turnover rate limits siRNA and microRNA efficacy. Mol Syst Biol 6:433. https://doi.org/10.1038/msb.2010.89
34. Sekar TV, Mohanram RK, Foygel K, Paulmurugan R (2013) Therapeutic evaluation of microRNAs by molecular imaging. Theranostics 3:964. https://doi.org/10.7150/thno.4928
35. Ameres SL, Horwich MD, Hung JH et al (2010) Target RNA-directed trimming and tailing of small silencing RNAs. Science 328:1534–1539. https://doi.org/10.1126/science.1187058
36. Kingston ER, Bartel DP (2019) Global analyses of the dynamics of mammalian microRNA metabolism. Genome Res 29:1777–1790. https://doi.org/10.1101/gr.251421.119
37. Fields C, Sheng P, Miller B et al (2019) Northern blot with IR fluorescent probes: strategies for probe preparation. Bio-protoc 9:e3219–e3219. https://doi.org/10.21769/BioProtoc.3219
38. Miller BR, Wei T, Fields CJ et al (2018) Near-infrared fluorescent northern blot. RNA 24:1871–1877. https://doi.org/10.1261/rna.068213.118
39. Kramer S (2017) Simultaneous detection of mRNA transcription and decay intermediates by dual colour single mRNA FISH on subcellular resolution. Nucleic Acids Res 45:e49–e49. https://doi.org/10.1093/nar/gkw1245

Check for updates

Chapter 5

RNA Labeling with Fluorophore

Joo-Hyung Lee

Abstract

Fluorescent labeling of RNA has become an essential technique for visualizing RNA molecules and interrogating their functions *in vitro* and *in vivo*. This chapter describes a detailed protocol for synthesizing fluorophore-labeled enhancer RNAs (eRNAs) through *in vitro* transcription with dye-modified nucleotides such as Cy3-UTP. The method allows efficient incorporation of fluorophores into RNA transcripts while maintaining their structural and functional integrity. Labeled RNAs can be used in a variety of downstream applications, including biochemical assays of RNA-protein interactions and phase separation studies. As demonstrated in prior studies, fluorophore-labeled eRNAs are sufficient to drive condensate formation with RNA-binding proteins such as YTHDC1, enabling direct visualization of RNA-mediated compartmentalization processes. The chapter also provides guidance on template preparation, RNA purification, and fluorescence validation, along with critical notes on optimizing labeling conditions and minimizing degradation. This method offers a robust and adaptable approach for producing fluorescent RNAs applicable to studies of RNA localization, nuclear architecture, and post-transcriptional gene regulation.

Keywords RNA labeling, eRNA, ncRNA, Phase-separation, Fluorophore-labeled RNA, m6A

1 Introduction

RNA labeling with fluorophores has become a powerful approach for visualizing RNA localization, monitoring RNA dynamics, and interrogating RNA-mediated molecular interactions both *in vitro* and *in vivo*. Among the diverse applications, fluorescent RNA probes have proven particularly useful in functional assays that rely on direct visualization of RNA behavior, such as biomolecular condensate formation, RNA trafficking, and interaction with RNA-binding proteins [1–3]. In recent years, growing interest in phase-separated biomolecular condensates has highlighted the functional roles of noncoding RNAs, including enhancer RNAs (eRNAs), as architectural components that facilitate condensate formation or modulate their biophysical properties [1]. In our previous study, we investigated the role of eRNAs in regulating phase separation by focusing on the *TFF1* enhancer RNA (*TFF1e*),

Je-Hyun Yoon (ed.), *RNA Labeling*, Methods in Molecular Biology, vol. 3047,
https://doi.org/10.1007/978-1-0716-5352-4_5,

a prototypical estrogen-responsive eRNA transcribed from a well-characterized enhancer locus in breast cancer cells [1]. We synthesized fluorescently labeled *TFF1e* transcripts by *in vitro* transcription using Cy3-UTP, and systematically incorporated N6-methyladenosine (m6A) into the RNA through partial substitution of ATP with m6ATP to mimic endogenous modification patterns. This approach enabled real-time imaging and biochemical interrogation of eRNA behavior in condensate-forming systems. Using purified YTHDC1, an m6A reader protein, we showed that labeled *TFF1e* transcripts could drive the formation of RNA-protein condensates under defined *in vitro* conditions [1]. Remarkably, m6A-modified *TFF1e* promoted larger and more robust condensate formation than unmodified RNA, indicating that m6A serves as a critical regulatory mark that enhances phase separation potential [1]. These results provided direct biochemical evidence that specific RNA modifications influence condensate assembly and offered a powerful demonstration of how fluorophore-conjugated RNAs can be used to dissect mechanistic aspects of nuclear organization and gene regulation.

This work builds on a broader body of research establishing that transcriptional coactivators associated with super-enhancers form phase-separated condensates to regulate gene expression [4]. It has also been shown that synthetic or endogenous RNAs can drive or stabilize the formation of these structures, as observed in studies using fluorophore-labeled long noncoding RNAs such as *NEAT1* in paraspeckle formation [5] or in reconstituted *in vitro* systems using synthetic RNA and purified RNA-binding proteins [2].

Together, these studies underscore the importance of site-specific and efficient RNA labeling strategies using fluorophores to enable mechanistic dissection of RNA functions beyond transcription, particularly in emerging fields such as phase separation biology, nuclear architecture, and post-transcriptional gene regulation. This chapter will provide detailed protocols for labeling RNA with fluorophores, including *in vitro* transcription with dye-modified nucleotides and post-synthetic conjugation methods, along with guidance for purifying and validating labeled RNA for downstream applications. An overview of the labeling workflow is shown in Fig. 1.

2 Materials

2.1 Preparation of DNA Templates for In Vitro Transcription

1. PCR primers (FWD: *GCTTAATACGACTCACTATAGGG (T7 promoter)* + CTAACTAGGCTCTTCCGGCCAAGTGAGTGAGGGGATGTGT (*TFF1e* FWD primer), REV: GCGGCCCTCTAGACTCGAGAAAGCAGCAGCAGATGAGATC (*TFF1e* REV primer)).
2. Genomic DNA from target cells.

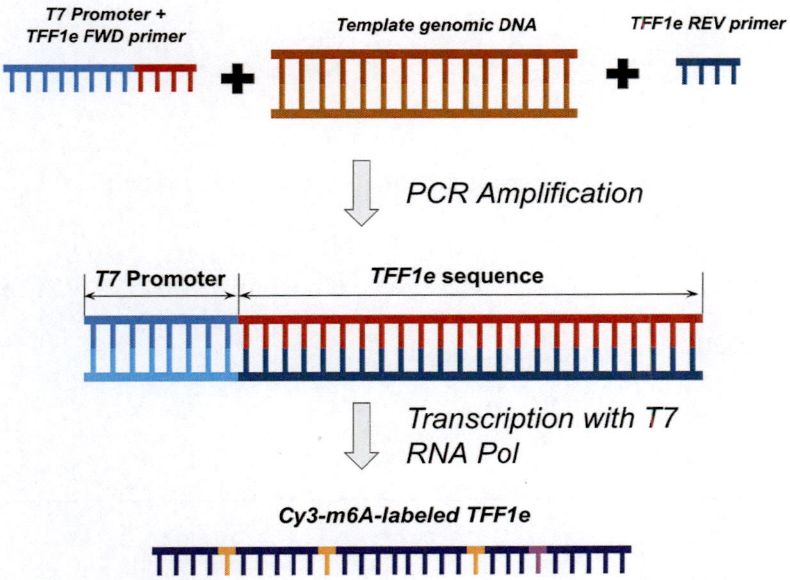

Fig. 1 Schematic overview of Cy3–m6A labeling of *TFF1* enhancer RNA (*TFF1e*). PCR amplification of the *TFF1e* template is performed using a forward primer containing a T7 promoter sequence and a reverse primer specific to the *TFF1e* locus. The resulting PCR product includes the T7 promoter upstream of the *TFF1e* sequence, which serves as a template for *in vitro* transcription. Transcription by T7 RNA polymerase in the presence of Cy3-UTP and m6ATP produces fluorescently labeled and m6A-modified *TFF1e* transcripts (Cy3–m6A–*TFF1e*), enabling visualization and functional interrogation of eRNA behavior in downstream assays. Created with Biorender.com

3. PCR-amplified enhancer DNA fragment (e.g., *TFF1* enhancer region) containing the T7 promoter sequence at the 5′ end.
4. Agarose gel and gel extraction kit.

2.2 In Vitro Transcription and RNA Labeling

1. MEGAscript™ T7 Transcription Kit (Invitrogen, Cat # AM1334).
2. Cy3-UTP (Enzo Life Sciences, Cat# ENZ-42505).
3. m6ATP (N6-Methyladenosine-5′-triphosphate, TriLink, Cat# N-1013-1) (optional for modified eRNA synthesis).
4. RNase-free water.
5. RNase Inhibitor.
6. DNase.
7. GlycoBlue.
8. 3 M Sodium acetate, pH 5.2.
9. 100% and 70% ethanol.
10. DEPC-treated water.

2.3 RNA Purification and Quality Control

1. TRIzol reagent.
2. Chloroform.
3. Isopropanol.
4. NanoDrop spectrophotometer.
5. Denaturing agarose or urea-PAGE system.

2.4 Phase Separation Assay Setup (Optional for Functional Validation)

1. Recombinant YTHDC1 protein.
2. Low-salt phase separation buffer: 20 mM HEPES (pH 7.5), 100 mM NaCl, 1 mM DTT, 5% glycerol.
3. Glass-bottom imaging dish.
4. Fluorescence microscope.

3 Methods

3.1 Preparation of DNA Templates for In Vitro Transcription

1. Amplify the enhancer RNA (eRNA) sequence from purified genomic DNA by PCR using primers that include a T7 promoter sequence at the 5′ end of the forward primer.
2. Purify the PCR product using a gel extraction.
3. Elute the DNA template in RNase-free water and quantify using NanoDrop. Typical yield should be ≥100 ng/μL.

3.2 In Vitro Transcription and Fluorophore Labeling

Set up the transcription reaction using the MEGAscript™ T7 kit according to the manufacturer's protocol with the following modifications:

1. Final reaction volume: 20 μL.
2. Use the following nucleotide concentrations: 5 mM ATP, CTP, GTP; 4 mM UTP + 1 mM Cy3-UTP (*see* **Note 1**).
3. For m6A-modified RNA, substitute a portion of ATP with m6ATP (up to 25%, *see* **Note 2**).
4. Add 1 μL of RNase Inhibitor to each reaction.
5. Incubate the reaction at 37°C for 4 h.
6. Treat the reaction with 1 μL Turbo DNase at 37°C for 15 min to remove the DNA template.
7. Stop the reaction by adding 80 μL of TRIzol reagent. Proceed to RNA purification.
8. All reactions should be conducted under low-light conditions to prevent fluorophore degradation (*see* **Note 3**).

3.3 RNA Purification and Precipitation

1. Add 20 μL of chloroform to the TRIzol mixture. Vortex for 15 s and incubate at room temperature for 2 min.

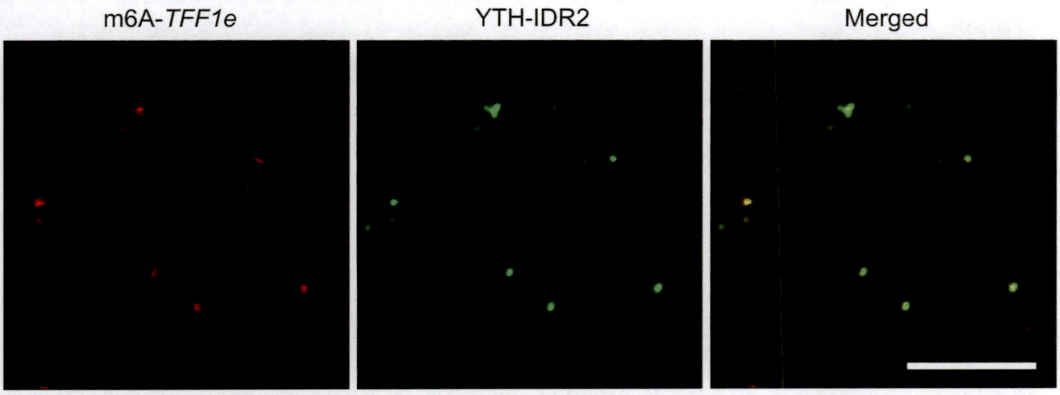

Fig. 2 Colocalization of m6A–TFF1e RNA with YTH-IDR2 *in vitro*. Confocal microscopy images showing Cy3-labeled m6A–*TFF1e* RNA (red, left), mClover-tagged YTH-IDR2 protein (green, middle), and their merged signals (right). Co-localization of RNA and protein indicates the recruitment of YTH-IDR2 to m6A-modified *TFF1e* RNA and the initiation of condensate formation. Scale bar, 10 μm

2. Centrifuge at 12,000 × *g* for 15 min at 4°C.
3. Transfer the upper aqueous phase (~40 μL) to a new tube.
4. Add 1 μL GlycoBlue, 4 μL 3 M sodium acetate, and 100 μL cold 100% ethanol.
5. Mix and precipitate the RNA at –20°C for ≥1 h (or overnight).
6. Centrifuge at 17,000 × *g* for 30 min at 4°C.
7. Wash the pellet with 1 mL of 70% ethanol, centrifuge briefly, air-dry, and resuspend in 20 μL RNase-free water (*see* **Note 4**).

3.4 Assessment of RNA Quality and Labeling Efficiency

1. Measure RNA concentration using NanoDrop.
2. Assess integrity by denaturing agarose gel or urea-PAGE if needed.
3. Confirm Cy3-label incorporation using a fluorescence spectrophotometer (excitation ~550 nm, emission ~570 nm).

3.5 (Optional) Phase Separation Assay with Labeled eRNA

1. Mix purified recombinant YTHDC1 protein (1 μM final) with Cy3-labeled *TFF1e* RNA (1 nM final) in phase separation buffer (10 mM HEPES, pH 7.4, 150 mM KCl, 3 mM $MgCl_2$, 0.01% NP-40, 10% PEG-8000).
2. Place the mixture onto a glass-bottom 35 mm dish. Incubate the sample in a 37°C humidified chamber for 30 min.
3. Acquire fluorescence images using a confocal microscope (see Fig. 2).
4. Analyze condensate particle size and morphology using ImageJ.

4 Notes

1. **Cy3-UTP incorporation efficiency and selection of dye:** Incorporating too much Cy3-UTP (e.g., >25% of total UTP) can reduce transcription efficiency and RNA yield. A final concentration of 10%–20% Cy3-UTP relative to total UTP is recommended for maintaining yield while achieving sufficient labeling intensity. Cy5, Alexa Fluor 488, or other dye-UTPs can be used following the same protocol, though incorporation rates and brightness vary. Always check dye compatibility with downstream assays (e.g., excitation/emission wavelengths, stability).
2. **Optimization of m6ATP ratio:** When using m6ATP to produce modified RNAs, excessive substitution (>25% of total ATP) may reduce transcription efficiency. Pilot experiments to titrate m6ATP incorporation are advised if functional assays depend on precise modification of stoichiometry.
3. **Fluorophore hydrolysis and storage:** Fluorescent dyes such as Cy3 are prone to hydrolysis and photobleaching. Labeled RNA should be stored in low-light conditions, aliquoted to avoid freeze–thaw cycles, and kept at –80°C in RNase-free water.
4. **Avoiding RNase contamination:** Use RNase-free consumables and reagents. Include RNase inhibitors in transcription and downstream steps, particularly during purification and phase separation assays.

References

1. Lee J-H, Wang R, Xiong F et al (2021) Enhancer RNA m6A methylation facilitates transcriptional condensate formation and gene activation. Mol Cell 81:3368–3385.e9. https://doi.org/10.1016/j.molcel.2021.07.024
2. Langdon EM, Qiu Y, Ghanbari Niaki A et al (2018) mRNA structure determines specificity of a polyQ-driven phase separation. Science 360:922–927. https://doi.org/10.1126/science.aar7432
3. Bratu DP, Cha B-J, Mhlanga MM et al (2003) Visualizing the distribution and transport of mRNAs in living cells. Proc. Natl. Acad. Sci. 100:13308–13313. https://doi.org/10.1073/pnas.2233244100
4. Sabari BR, Dall'Agnese A, Boija A et al (2018) Coactivator condensation at super-enhancers links phase separation and gene control. Science 361:eaar3958. https://doi.org/10.1126/science.aar3958
5. Yamazaki T, Souquere S, Chujo T et al (2018) Functional domains of NEAT1 architectural incRNA induce paraspeckle assembly through phase separation. Mol Cell 70:1038–1053.e7. https://doi.org/10.1016/j.molcel.2018.05.019

Check for updates

Chapter 6

Labeling of Amine-Modified RNA

Sujin Kim and Myung Hyun Jo

Abstract

RNA labeling enables high-contrast visualization and quantitative measurements. This chapter describes practical guidance for fluorescent dye labeling of RNA via the NHS ester–amine coupling. We detail reaction setup, cleanup by ethanol precipitation, and evaluation of labeling efficiency by absorbance spectrum analysis or by high-performance liquid chromatography (HPLC). Notes summarize common failure modes and remedies, and RNase-free practice is emphasized throughout. Although the workflow demonstrates fluorescent labeling of 3′-amine–modified RNA, the protocol is broadly applicable to 5′-end or internal amine modifications and can also be used to introduce functional groups such as DBCO, used in click chemistry reactions.

Keywords RNA labeling, End labeling (5′/3′), Internal labeling, Fluorophore conjugation, Labeling efficiency, Amine-NHS ester reaction

1 Introduction

RNA can be labeled at defined sites to support imaging, quantification, and interaction analysis [1–4]. Common labels include fluorophores, radioisotopes, affinity tags (e.g., biotin), and bio-orthogonal reactive groups (e.g., azide/alkyne) for click chemistry. Labels can be introduced at the 5′ or 3′ termini or placed internally, and the choice of modification and linker can affect signal intensity and labeling efficiency.

The RNA labeling protocol is not fundamentally different from DNA labeling, except that strict RNase control is required. Amine-NHS esters, thiol-maleimide, and Strain-Promoted Azide-Alkyne Cycloaddition (SPAAC) options are all applicable to nucleic acids [3–8]. Among these, amine-NHS ester is a cost-effective and versatile option given the broad commercial availability of amine-modified oligos (3′, 5′, or internal U) and NHS ester dyes/crosslinkers. Thiol-maleimide conjugation is also applicable but, unlike amine-based labeling, it may alter photostability, requiring careful consideration before use [9, 10]. Accordingly, this chapter presents the workflow

Je-Hyun Yoon (ed.), *RNA Labeling*, Methods in Molecular Biology, vol. 3047,
https://doi.org/10.1007/978-1-0716-5352-4_6,

in parallel with common DNA-labeling practices while emphasizing RNA-specific considerations that can affect yield.

Labeling site selection should consider experimental objectives, RNA length and structure, and compatibility with downstream steps. Complete removal of unreacted dyes is important to estimate labeling efficiency. Optionally, high-performance liquid chromatography (HPLC) can be used to distinguish labeled RNA from unlabeled RNA or purify the labeled RNA, while gel electrophoresis (native PAGE or denaturing urea-PAGE) provides complementary information on product integrity. Strict RNase-free practice throughout all steps is critical to preserve RNA integrity [11–14].

This chapter presents 3′-amine-modified RNA labeled by sulfo-Cyanine 3 (Cy3) via amine-NHS ester conjugation as the working example. Step-by-step parameters and troubleshooting are detailed in Methods and Notes.

2 Materials

2.1 Reagents

1. 3′-amine-modified target RNA (e.g., IDT modification code /3AmMO/; primary amine introduced at the 3′ terminus) [15]. Prepare 1- or 0.5-mM stocks in nuclease-free water; store at −20°C.
2. NHS-ester fluorophore (e.g., sulfo-Cy3 NHS ester). Prepare a 20 mM stock in anhydrous DMSO to protect from moisture and subsequent hydrolysis; protect from light (*Lumiprobe, Hunt Valley, MD, USA; cat. no. 11320*) [16].
3. Anhydrous dimethyl sulfoxide (DMSO) for dye preparation and dye dissolving.
4. 100% Ethanol (ethyl alcohol), molecular-biology grade for precipitation and pellet wash. Prepare 80% (v/v) fresh using RNase-free water for pellet wash.
5. 200 mM EPPS, pH 8.5 for pH buffer.
6. 5 M NaCl.
7. Nuclease-free water.
8. Storage buffer: RNase-free water or PBS for short-term storage.
9. (Optional) Moisture absorber (e.g., Drierite) to maintain anhydrous DMSO for dye stock preparation.
10. (Optional) 100 mM Sodium bicarbonate, pH 8.3–8.5 in case EPPS, pH 8.5 is not available.
11. (Optional) TE50: 10 mM Tris-HCl, pH 8.0, 50 mM NaCl, and 0.1 mM EDTA can be used for long-term storage. Avoid Tris-containing buffers if further reactions are expected.

12. (Optional) Glycogen (20 g/L) as a co-precipitant to improve yield and pellet visibility during ethanol precipitation.
13. (Optional, for HPLC analysis) 50 mM triethylamine acetate (TEAA) as Buffer A and acetonitrile as Buffer B.
14. (Optional, for PAGE gel electrophoresis analysis) 40% acrylamide:bisacrylamide (29:1), 10× TAE (native) or TBE (denaturing), urea (for denaturing gel), TEMED, 10% ammonium persulfate (APS).

2.2 Equipment

1. Microcentrifuge (≥16,000 × *g*, 4°C).
2. Spectrophotometer for A_{260} and dye A_{max}.
3. (Optional) HPLC with absorbance detection (260 nm plus dye A_{max}); RP-C18 column recommended (e.g., *Agilent 1260 Infinity system; or equivalent*).
4. (Optional) PAGE setup (native or denaturing) and a fluorescence-capable imager (e.g., *iBright FL series, Thermo Fisher Scientific; or equivalent*).

3 Methods

The workflow comprises (i) amine–NHS conjugation, (ii) ethanol precipitation, and (iii) evaluation of labeling efficiency by absorbance spectroscopy, and optional HPLC or PAGE analysis. All steps must be performed under strict RNase-free conditions (*see* **Note 1**).

3.1 Amine–NHS Conjugation (3′ End)

1. Prepare reaction mixture for each 50 μL reaction as shown below, adding the components in the following order: RNA, RNase-free water, EPPS, pH 8.5, and finally the NHS ester dye (*see* **Note 2**).

Component	Final conc.
RNA-3′Am	100 μM
NHS ester dye	1 mM
EPPS pH 8.5	20 mM
RNase-free water	Up to 50 μL
Total volume	50 μL

Freshly prepared sodium bicarbonate (~pH 8.3–8.5, 100 mM) is also suitable instead of EPPS; Add the pH buffer just before the reaction to minimize NHS-ester hydrolysis. Avoid Tris or other reagents that contain primary amines (*see* **Note 3**).
Prepare the NHS-ester conjugated dye or crosslinker stock in anhydrous DMSO and adjust the reaction mix to a final < 20% (v/v) DMSO; for highly hydrophobic dyes, increase up to 50% (v/v) (*see* **Notes 4, 5, 6, and 12**).

2. Mix gently and protect from light. Verify that the dye remains fully dissolved and that no precipitation has occurred (can be checked by a brief spin-down).
3. Incubate 1 hr at room temperature (20–25 °C) or ≥ 4 h at 4 °C, protected from light.

3.2 Ethanol Precipitation

1. Add 10 μL of 5 M NaCl and 140 μL of 100% ethanol to the 50 μL reaction (final 70% ethanol). Incubate 10 min on ice; extend the incubation as needed.
2. Centrifuge 10 min at 16,000 × g, 4°C. Carefully remove the supernatant (retain the supernatant if pellet is faint; *see* **Note 7**).
3. Resuspend the pellet in 50 μL RNase-free water.
4. Second precipitation to remove free dye completely: add 10 μL of 5 M NaCl, 140 μL of 100% ethanol, and incubate 10 min on ice.
5. Centrifuge 10 min at 16,000 × *g*, 4°C. Remove the supernatant. If the supernatant is not clear, repeat until free dye is no longer visible.
6. Wash the pellet three times: 200 μL 100% ethanol → discard; 200 μL 80% ethanol → discard; 200 μL 100% ethanol → discard.
7. Briefly spin down the tube and remove the residual ethanol completely.
8. Dry the pellet at room temperature in the dark (or under a mild vacuum for ~5 min). Take care not to overdry the pellet.
9. Resuspend the pellet in RNA storage buffer (e.g., PBS or TE50). Avoid TE50 or amine-containing buffers (e.g., Tris) if there is a chance that labeling will need to be repeated because of low efficiency.

If the labeling efficiency is too low, perform second labeling (re-labeling) (*see* **Notes 8 and 9**). If the labeling efficiency is higher than 100%, repeat the free dye removal and re-measure the efficiency (*see* **Note 10**).

3.3 Evaluation of Labeling

Labeling efficiency can be determined by absorbance spectroscopy. The concentration of RNA and the dye can be calculated using their respective absorbance values at a given wavelength and extinction coefficients. For further analysis—such as purification or RNA integrity assessment—optional techniques including HPLC and PAGE may be used.

3.3.1 Labeling Efficiency by Absorbance Spectroscopy

1. Measure A_{260} and $A_{\lambda dye}$ (e.g., sulfo-Cy3 A_{max} = 550 nm) using spectrophotometer (e.g. Nanodrop, 17, 18).
2. Correct for dye absorbance at 260 nm:

$$A_{260,corr} = A_{260} - CF_{260} \times A_{\lambda dye}$$

where CF_{260} is the vendor-provided correction factor (e.g., Cy3 $CF_{260} = 0.03$).

3. Compute concentrations:

$$[\text{RNA}] = \frac{A_{260,corr}}{e_{260,RNA}}; [\text{dye}] = A_{\lambda dye}/e_{\lambda dye}.$$

4. Labeling efficiency = [dye]/[RNA]. Document path length and extinction coefficients used.

A representative absorbance spectrum of Cy3-labeled RNA is shown in Fig. 1, demonstrating distinct peaks at 260 nm (RNA) and ~550 nm (Cy3). A representative calculation example is summarized in Table 1, illustrating the determination of RNA concentration, dye concentration, and labeling efficiency.

3.3.2 HPLC (Optional; Recommended for Clear Distinction and Purification)

HPLC is recommended for a clear distinction between unlabeled RNA and labeled RNA. RNA, Cy3, RNA-Cy3 was analyze by ion-pair reversed-phase HPLC on a Poroshell 120 EC-C18 column (4.6 × 100 mm, 2.7 μm). Absorbance was monitored at 260 nm and dye A_{max} (e.g., 550 nm for Cy3), as shown in Fig. 2. Labeled RNA typically elutes later than unlabeled RNA. Integrate peak areas to estimate labeling fraction.

3.3.3 PAGE (Optional; Recommended for Validation of Product Integrity)

Native PAGE (TAE) can visualize fluorescence between labeled and unlabeled RNA; include unlabeled and dye-only controls. Denaturing urea-PAGE (TBE + 8 M urea) is recommended when size resolution or removal of secondary structure is required [19, 20].

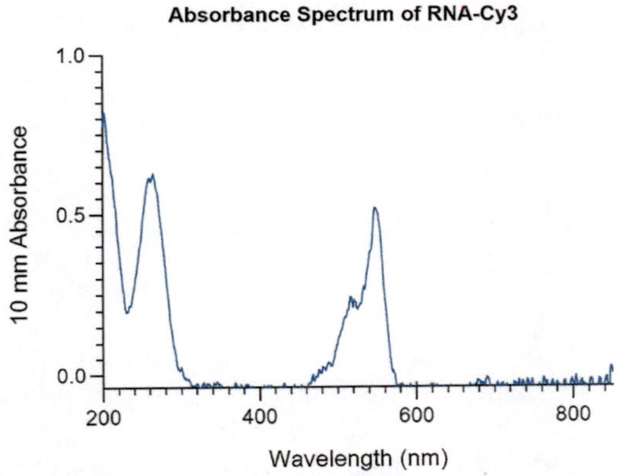

Fig. 1 Absorbance spectrum (Nanodrop; 220–850 nm). Absorbance spectrum of Cy3-labeled RNA showing RNA peak at 260 nm and the dye peak at ~550 nm

Table 1
Example labeling efficiency calculation worksheet using $A_{260,corr}$, A_{550}, and extinction coefficients

A_{260}_obs	0.61
A_{550}_obs	0.51
CF_{260}	0.03
A_{260}_corr	0.6
ε_{260}_RNA_(M^{-1} cm^{-1})	134,400
ε_{550}_dye_(M^{-1} cm^{-1})	150,000
RNA conc. (μM)	4.4
Dye conc. (μM)	3.4
Labeling efficiency ([Dye]/[RNA])	0.77

The extinction coefficient of RNA is sequence-specific (*see* **Note 11**), and the extinction coefficient and correction factor of dye are dye-specific. The sample was diluted 20-fold for measurement

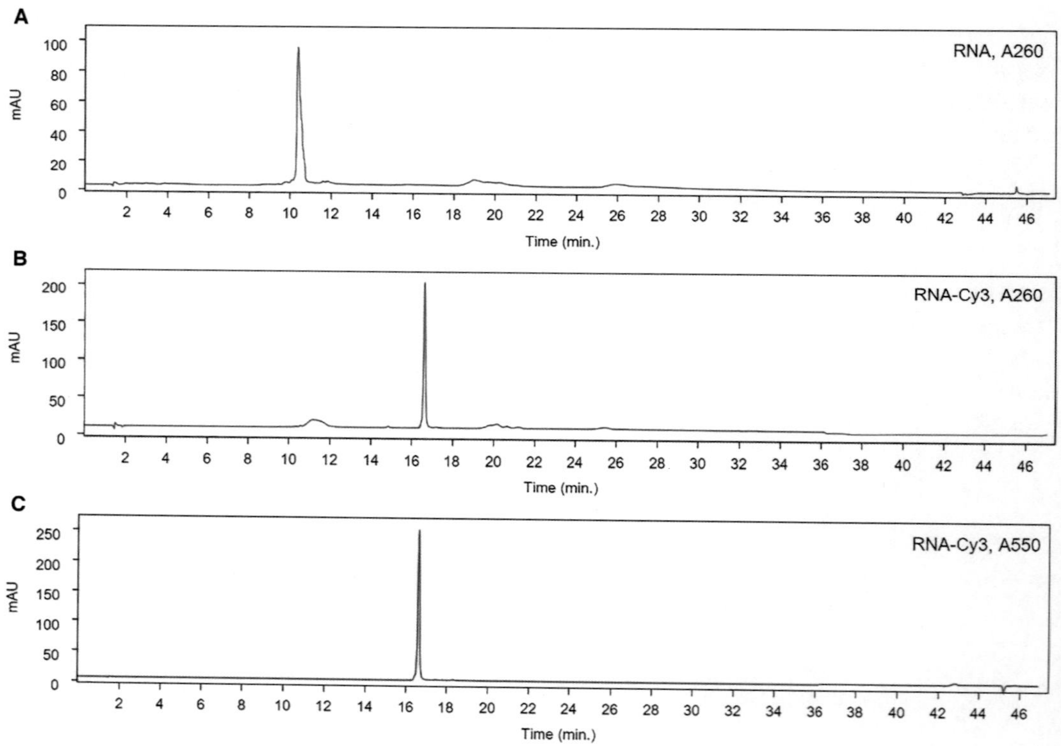

Fig. 2 Absorbance chromatograms of unlabeled RNA, Cy3, and Cy3-labeled RNA. (**A**) Unlabeled RNA with a sharp peak at 10.4 min (260 nm). (**B** and **C**) Cy3-labeled RNA with a peak at 16.6 min at both A260 and A550. Unreacted Cy3 peak was at ~23 min (550 nm) (not presented here). The analysis was held with a flow of 1 mL/min, using 50 mM TEAA as Buffer A and acetonitrile as Buffer B. Samples were dissolved in 50 mM TEAA. A gradient was 5% B (0–5 min), 5 to 45%). Detection employed a DAD monitoring at 260 nm and dye A_{max} (550 nm for Cy3)

4 Notes

1. RNase control: Use dedicated pipettes and RNase-free plastics/glassware; change gloves frequently; decontaminate work surfaces and tools with ethanol [14].
2. EPPS, pH 8.5 sets the mixture to the reaction pH, where NHS esters are reactive but susceptible to hydrolysis. Add the NHS ester dye last so it can immediately encounter RNA, minimizing hydrolysis and maximizing the yield.
3. Maintain the pH tightly within 8.3–8.5 during the reaction [21].
4. Prepare fresh dye aliquots. To minimize freeze-thaw cycles, it is recommended to aliquot in small volume. For stock preparation, dissolve the powder in anhydrous DMSO and minimize handling time. Store at −20°C or −80°C. If possible, vacuum-packaging can help preserve dye quality.
5. Anhydrous DMSO used for dye preparation and for dissolving highly hydrophobic dyes must be kept dry. Use moisture absorber (e.g., Drierite) to prevent moisture uptake. For a simple setup, place the moisture absorber at the bottom of 50 mL tube, insert one Kimwipe as a spacer to keep the moisture absorber separate, set DMSO aliquots on top, and seal the cap tightly. Alternatively, vacuum-seal the DMSO aliquots for storage.
6. If solubility of the dye limits efficiency, add DMSO up to 50% (v/v).
7. Pellet recovery: If the pellet is faint or absent, pool supernatants, add 1 μL glycogen (20 g/L), extend the ice incubation, and repeat centrifugation [22].
8. In case the labeling efficiency is low (10%–70%), perform a second labeling (re-labeling) with additional fresh NHS ester dye.
9. Labeling efficiency <10% indicates a failed labeling. Verify dye freshness, dye solubility, pH 8.3–8.5, reaction time, and amine contamination.
10. If the labeling efficiency is higher than 100%, it indicates free dye was not removed completely. Repeat ethanol precipitation to remove residual dye.
11. The extinction coefficient of the RNA at 260 nm depends on how many A, C, G, and U it contains, which makes it sequence-specific.

12. NHS-linked bio-orthogonal handles (azide/alkyne) enable subsequent CuAAC or SPAAC conjugation; balance reaction efficiency against copper cytotoxicity for cellular applications.

Acknowledgments

This work was supported by Korea University and the National Research Foundation of Korea (NRF) grant funded by the Korean government (RS-2025-00517258).

References

1. Braselmann E, Rathbun C, Richards EM, Palmer AE (2020) Illuminating RNA biology: tools for imaging RNA in live mammalian cells. Cell Chem Biol 27(8):891–903
2. Williams CG et al (2022) An introduction to spatial transcriptomics for biomedical research. Genome Med 14(1):68
3. Raj A et al (2008) Imaging individual mRNA molecules using multiple singly labeled probes. Nature Methods 5(10):877–879
4. Chen KH et al (2015) Spatially resolved, highly multiplexed RNA profiling in single cells. Science 348(6233):aaa6090
5. Hayes OG, Partridge BE, Mirkin CA (2021) Encoding hierarchical assembly pathways of proteins with DNA. Proc Natl Acad Sci 118(40):e2106808118
6. Huang X et al (2021) DNA scaffolds enable efficient and tunable functionalization of biomaterials for immune cell modulation. Nat Nanotechnol 16(2):214–223
7. Pan L et al (2020) DNA-mediated assembly of multispecific antibodies for T cell engaging and tumor killing. Adv. Sci. 7(2):1900973
8. Baranda Pellejero L, Nijenhuis MA, Ricci F, Gothelf KV (2023) Protein-templated reactions using DNA-antibody conjugates. Small 19(13):2200971
9. Gidi Y et al (2020) Unifying mechanism for thiol-induced photoswitching and photostability of cyanine dyes. J Am Chem Soc 142(29):12681–12689
10. Zhang Y et al (2023) General strategy to improve the photon budget of thiol-conjugated cyanine dyes. J Am Chem Soc 145(7):4187–4198
11. Li S et al (2024) RNA quality score evaluation: a preliminary study of RNA integrity number (RIN) and RNA integrity and quality number (RNA IQ). Forensic Sci Int 357:111976
12. Fleige S, Pfaffl MW (2006) RNA integrity and the effect on the real-time qRT-PCR performance. Mol Aspects Med 27(2-3):126–139
13. Gallego Romero I, Pai AA, Tung J, Gilad Y (2014) RNA-seq: impact of RNA degradation on transcript quantification. BMC Biol 12(1):42
14. Green MR, Sambrook J (2019) How to win the battle with RNase. Cold Spring Harbor Protocols 2019(2):top101857
15. Integrated DNA Technologies (2019) 3′ Amino modifier (/3AmMO/): product page and technical details. In. IDT DNA. Integrated DNA Technologies, Coralville, IA, USA
16. Lumiprobe (2025) Sulfo-Cyanine3 NHS ester
17. Thermo Fisher, S (2025) Accurate oligonucleotide quantification with nanodrop spectrophotometers. Thermo Fisher Scientific, Waltham, MA, USA
18. Thermo Fisher, S (2018) A theoretical and practical guide for spectrophotometric determination of protein concentrations at 280 nm. In: NanoDrop One/OneC and Multiskan Sky spectrophotometers. Thermo Fisher Scientific, Waltham, MA, USA
19. Green MR, Sambrook J (2021) Separation of RNA according to size: electrophoresis of RNA through denaturing urea polyacrylamide gels. Cold Spring Harbor Protocols 2021(1):prot101766
20. Thermo Fisher S (2012) Denaturing polyacrylamide/urea gel electrophoresis. Thermo Fisher Scientific, Waltham, MA, USA
21. Trabjerg E, Keller A, Leitner A (2021) pH dependence of succinimide-ester-based protein cross-linking for structural mass spectrometry applications. ACS Meas Sci Au 2(2):132–138
22. Green MR, Sambrook J (2020) Precipitation of RNA with ethanol. Cold Spring Harbor Protocols (3):pdb.prot101717. https://doi.org/10.1101/pdb.prot101717

Part II

Methods for Study of RNA Localization

Chapter 7

smFISH: Quantifying and Mapping RNA at the Single-Molecule Level

Noah S. Helton and Stephanie L. Moon

Abstract

RNA localization regulates spatial and temporal control of protein synthesis, influencing processes such as development, neuronal signaling, cell migration, and stress responses. Single-molecule fluorescence *in situ* hybridization (smFISH) enables direct visualization and quantification of individual transcripts within cells, providing both absolute counts and subcellular localization of RNAs. Here, we present a detailed, step-by-step protocol for smFISH in adherent mammalian cells, including probe design and labeling, sample preparation, hybridization, imaging, and analysis. This guide offers a reproducible workflow that facilitates the study of RNA localization and expression at single-molecule resolution.

Keywords smFISH, RNA localization

1 Introduction

RNA localization can dictate gene expression and is a critical form of post-transcriptional gene regulation. Messenger RNAs (mRNAs) are synthesized in the nucleus and must be exported to the cytoplasm to translate into proteins. After exiting the nucleus, mRNAs can be further compartmentalized to regulate local translation and underlie many diverse biological processes. For example, in neuronal cells, mRNA targeting to axons and dendrites is critical for plasticity, learning, and responding to stress [1–4]. Further, many studies have demonstrated that mRNA localization is important for proper *Drosophila* development, best exemplified by *bicoid* and *oskar* RNAs that help establish polarity in the oocyte via local translation on anterior and posterior poles [5–12]. RNA localization also contributes to the cellular stress response, as constitutively expressed transcripts can be sequestered into stress granules, while translating stress-induced gene mRNAs are inhibited from localizing to stress granules [13, 14]. Misregulation of RNA localization is associated with human diseases, including neurodevelopmental disorders, neurodegeneration, and

Je-Hyun Yoon (ed.), *RNA Labeling*, Methods in Molecular Biology, vol. 3047,
https://doi.org/10.1007/978-1-0716-5352-4_7,

cancer [15–18]; thus, there is a critical need to have techniques to visualize RNA localization. Indeed, single-molecule fluorescence *in situ* hybridization (smFISH) enables the direct visualization and quantification of individual transcripts within cells, revealing both their subcellular positioning and abundance.

smFISH builds on early fluorescence *in situ* hybridization methods and was first described in the late 1990's as a way to visualize single transcripts using multiple short, fluorescently labeled DNA probes hybridized along a target RNA [19, 20]. This approach provides absolute transcript counts as well as precise spatial information within fixed cells and tissues. smFISH involves fixation and permeabilization of samples, probe hybridization, and probe detection with fluorescence microscopy (Fig. 1). Immunofluorescence or imaging of fluorescently labeled proteins in the same samples permits investigation of RNA colocalization with proteins and/or compartmentalization in different cell types or states. Compared to bulk transcriptomic methods, smFISH offers direct spatial resolution at the single-cell level, making it a cornerstone technique for dissecting RNA localization, cellular heterogeneity, and regulatory mechanisms of gene expression.

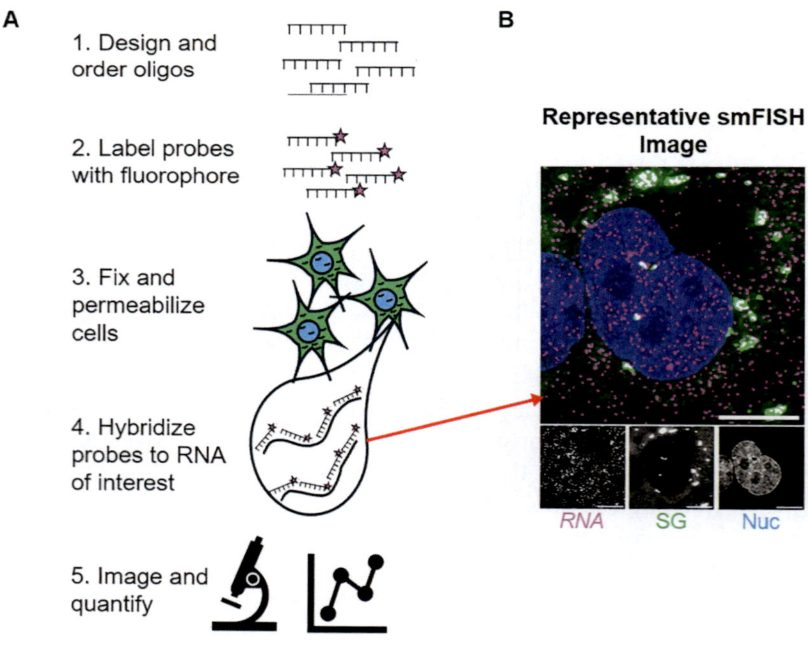

Fig. 1 Overview of single-molecule fluorescence *in situ* hybridization (smFISH) protocol. (**A**) Summary of step-by-step protocol to perform smFISH in adherent mammalian cells. (**B**) Representative smFISH image. U-2 OS cells were stressed for 45 min with 250 μM arsenite prior to fixation. Magenta represents RNA hybridized with FISH probes labeled with ATTO-647. Green is the stress granule (SG) marker, G3BP1, tagged with GFP, and the blue is nuclei stained with Hoechst #33342. Images are max intensity projections deconvolved using the Richardson-Lucy deconvolution. Scalebar = 10 μM

In this chapter, we provide a practical guide for performing smFISH in adherent mammalian cells, covering all stages from probe design and labeling to sample preparation, hybridization, and imaging. We detail experimental considerations for fixation and permeabilization that preserve RNA and cellular structure, as well as strategies for optimizing probe hybridization. Finally, we describe approaches for imaging and quantitative analysis, including methods for identifying and counting individual transcripts. Together, these procedures provide a reproducible workflow for visualizing and quantifying RNA localization at single-molecule resolution.

2 Materials

2.1 Probe Labeling

1. Oligonucleotides to RNA target.
2. Terminal deoxynucleotidyl Transferase (TdT) Enzyme and Buffer (Thermo Scientific).
3. ddUTP-dye (such as ddUTP-ATTO565, ddUTP-ATTO633, ddUTP-ATTO647N, or ddUTP-Cy5; Jena Biosciences).
4. RNase-free water.
5. Thermocycler.
6. Spectrophotometer.
7. PCR-strip tubes.

2.2 Performing smFISH on Adherent Mammalian Cells with Optional Sequential Immunofluorescence

1. Adherent tissue culture cells, such as U-2 OS cells.
2. Humidified cell culture incubator with 5% CO_2.
3. Cell culture media.
4. Phosphate-buffered saline (PBS).
5. 4% paraformaldehyde (PFA) solution diluted in PBS.
6. 96-well glass-bottom plates (#1.5H) or equivalent sterilized coverslip in a 12-well cell culture dish.
7. 12-well culture dishes (if using coverslips).
8. Wide-field fluorescence microscope with a high numerical aperture (>1.3) and a 60–100× oil-immersion objective.
9. 20× SSC:
   a. 175.3 g of NaCl (3 M).
   b. 88.2 g of sodium citrate (0.3 M).
   c. To 1 L with Molecular Grade H_2O.
10. Deionized formamide.
11. Wash Buffer A (WBA):
    a. 15 mL 20× nuclease-free SSC.
    b. 120 mL Molecular Grade H_2O.

12. Wash Buffer B (WBB):
    a. 15 mL 20× nuclease-free SSC.
    b. 135 mL nuclease-free water.
13. 0.1% Triton-× 100:
    a. 100 μL of 10% Triton-× 100.
    b. 9.9 mL of PBS.
14. RNase inhibitor (Ribolock).
15. Reagents for immunofluorescence (**Optional**).
    a. Primary and secondary antibodies.
    b. Bovine serum albumin (BSA).
    c. Antibody Dilution (Abdil) Buffer in PBS:
        i. 0.5% Triton-× 100.
        ii. 3% BSA.
    d. 3% BSA in PBS.
16. NucBlue with Hoechst #33342 (**Optional nuclear stain**).
17. Hybridization buffer
    a. 2 mL 50% dextran sulfate.
    b. 1 mL deionized formamide.
    c. 1 mL 20× nuclease-free SSC.
    d. 6 mL nuclease-free H_2O.

2.3 Software for Analysis

1. Python V3.6 or greater with BigFISH [21].
2. ImageJ/FIJI.

3 Methods

3.1 Probe Design

Properly designing probes is essential for the specificity of smFISH. Ensure you are using the primary transcript of interest (keeping in mind splicing isoforms), use Stellaris's designer to match the organism of interest, and leave masking level as high as possible to reduce potential off-target hybridization. Increasing the number of probes will increase the signal-to-noise ratio of your smFISH spots. Following the recommended spacing and probe size is also critical for specificity and proper hybridization.

1. Design DNA probes using Stellaris Probe Designer: https://oligos.biosearchtech.com/stellaris-designer
2. Leave the masking level at 5. Lower masking levels increase the chances of non-specific binding.
3. Design 25–48 probes for RNA target.

4. Select probe length of 20 nt with 2 nt spacing.
5. Insert your target sequence and hit Design Probes.
6. Order the probes (e.g., from IDT) either desiccated or resuspended in RNase-free water at 200 μM concentration.
7. Make a master mix of probes by adding 5 μL of each probe to a tube.

3.2 Probe Labeling and Clean Up

This protocol was adapted from [22, 23], which provides a more detailed explanation of the labeling reaction.

1. To make the labeling reaction, mix the following reagents in a PCR strip tube
   a. 5 μL of oligonucleotide probe master mix.
   b. 2 μL of ddUTP-fluorophore (1 mM).
   c. 3 μL of 5× TdT buffer.
   d. 1 μL TdT (30 U).
   e. 4 μL nuclease-free water.
2. Incubate reaction in a thermocycler with a heated lid for 20–24 h at 37°C.
3. Adjust the volume of reaction to 50 μL with nuclease-free water.
4. Use Zymogen's Oligo cClean and concentrate kit to clean up after probe labeling and elute in 25 μL (*see* **Note 1**).
5. Measure the absorbance of labeled probes at 260 nm and the wavelength of dye used (*see* **Note 2**) using a spectrophotometer. Calculate probe concentration and labeling efficiency using the following equations:

$$[DNA] = \frac{(A_{260} \cdot 33) \cdot 0.001}{(L \cdot 308.97) + 79.97} \cdot 10^6$$

$$[Fluor] = \frac{A_{Fluor}}{\varepsilon \cdot b} \cdot 10^6$$

$$Labeling\ Efficiency = \frac{[Fluor]}{[DNA]}$$

Here, A260 is the concentration absorbance of DNA at 260 nm multiplied by 33 to obtain ng/μL of DNA. This number is divided by the nucleotide length (L) multiplied by the average nucleotide molecular weight plus the terminal phosphate to calculate DNA molarity (μM). The fluorophore concentration in μM is calculated using Beer-Lambert's law using the extinction coefficient of the fluorophore (ε) and a path length of 1 centimeter (could vary depending on the spectrophotometer being used). The labeling efficiency is determined by dividing the concentration of labeled probes (Fluor) by the total DNA concentration.

6. If labeling efficiency is below 70%, repeat steps 1 to 5, replacing oligonucleotide probe master mix with your eluate from step 4.
7. Dilute probes to 12.5 μM in nuclease-free water and store protected from light at −20°C.

3.3 Performing smFISH on Adherent Mammalian Cells (with Sequential Immunofluorescence)

All reagents and solutions should be sterile, and when possible, RNase-free.

1. Plate adherent cells in a 96-well glass-bottom plate or on top of a sterilized coverslip in a 12-well cell culture dish such that they reach ~70%–80% confluency by the time of the experiment (*see* **Notes 3 and 4**).
2. After your experiment, wash the cells by adding 200 μL (96-well plate) or 250 μL (12-well dish) of PBS to each well.
3. Remove the PBS. In a fume hood, add 200 μL (96-well plate) or 250 μL (12-well dish) of 4% PFA to fix the cells (*see* **Note 5**).
4. Incubate at room temperature for 10 min.
5. Remove PFA into the appropriate waste container.
6. Wash two times with 200 μL (96-well plate) or 250 μL (12-well dish) of PBS.
7. **DECISION POINT**: If performing immunofluorescence, follow these next few steps. If no immunofluorescence is necessary for your samples (e.g., GFP-tagged cell lines), move on to permeabilization with Triton at step 8 (*see* **Note 6**).
   a. Permeabilize in 100 μL of Abdil Buffer + 0.4 μL Ribolock (96-well plate) or 250 μL Abdil Buffer + 1 uL (12-well dish) for 10 min at room temperature (*see* **Note 7**).
   b. Wash once in 200 μL (96-well plate) or 250 μL (12-well dish) PBS.
   c. Incubate for 1 hour with primary antibody of interest in 100 μL 3% BSA + 0.4 μL of Ribolock (96-well plate) or 250 μL of 3% BSA + 1 μL of Ribolock (12-well dish).
   d. Wash three times in 200 μL (96-well plate) or 250 μL (12-well dish) PBS.
   e. Incubate for 1 h with secondary antibody of interest in 100 μL of 3% BSA + 0.4 μL of Ribolock (96-well plate) or 250 μL 3% BSA + 1 μL Ribolock (12-well dish).

f. Wash three times in 200 μL (96-well plate) or 250 μL (12-well dish) PBS.
g. Fix cells again with 200 μL (96-well plate) or 250 μL (12-well dish) of 4% PFA for 10 min at room temperature.
h. Wash two more times in 200 μL (96-well plate) or 250 μL (12-well dish) PBS, and continue to WBA wash step (Step 10).

8. To permeabilize cells if no immunofluorescence is necessary, incubate cells with 100 μL of Triton-X 100 (0.1%) + 0.4 μL of RNase inhibitor (Ribolock) (96-well plate) or 500 μL 0.1% Triton-X 100 + 2 μL Ribolock (12-well dish).
9. Incubate at room temperature for 5 min.
10. To wash, incubate with 200 μL (96-well plate) or 1 mL (12-well dish) WBA + 10% formamide for 5 min at room temperature.
11. Remove WBA into a designated waste container, as formamide is toxic.
12. Hybridize smFISH probes to RNA of interest by adding 75 μL (96-well plate) or 100 μL (12-well dish) of hybridization buffer + 1 μL of purified probes (12.5 μM) to each well/coverslip. If working with coverslips, flip the cell-side down onto a droplet of hybridization solution on parafilm over wet paper towels in a petri dish.
13. Wrap the 96-well plate or petri dish in parafilm and incubate at 37°C for 16–20 h protected from light.
14. Wash once in 200 μL (96-well plate) or 1 mL (for coverslips placed into a new 12-well dish) of WBA + 10% formamide + NucBlue dye (1 drop of NucBlue for 1 mL of WBA; *see* **Note 8**). If working with a coverslip, remove the coverslip from the incubation chamber into a 12-well dish.
15. Incubate for 30 min at 37 °C protected from light.
16. Remove WBA and wash once more in 200 μL (96-well plate) or 1 mL (12-well dish) WBA + 10% formamide.
17. Incubate for 30 min at 37°C protected from light.
18. Remove WBA, add 200 μL (96-well plate) or 1 mL (12-well dish) of WBB, and continue to imaging (*see* **Note 9**).
19. If working with coverslips, mount the coverslip onto a slide with the cell-side down and seal with clear nail polish.

3.4 Image Acquisition and Analysis

Optimized imaging parameters and proper analysis tools are essential for high-quality smFISH data. We recommend creating a custom pipeline depending on your use case with BigFISH/FISH-Quant functions built in to reduce bias and increase reproducibility. These use cases could include subcellular localization of the RNA, cellular heterogeneity, or total RNA counts.

1. To acquire high-resolution images of single FISH spots, we recommend acquiring images in either widefield or HILO (Highly Inclined Thin Illumination) on a widefield microscope with a 100X oil objective.
2. Determine imaging parameters optimal for your experiment (e.g., exposure time and laser intensity).
3. Next, determine the Z slices necessary to acquire the full cell volume, 13–15 Z-slices at 0.2 μm intervals for U-2 OS cells are sufficient, if necessary.
4. Collect three images in each well for analysis.
5. For image analysis, we highly recommend using BigFISH/FISH-Quant pipeline to accurately count single-molecule FISH spots, segment cells and nuclei, and determine subcellular localization (*see* **Note 10**).
   a. Import raw images into the BigFISH pipeline and recommend segmentation of cells and nuclei using a pre-trained U-net model available from BigFISH using TensorFlow.
   b. Select well-segmented cells and perform downstream smFISH analysis within those cells.
   c. Identify threshold values for smFISH spots that represent your images. FISH spots are distinct single points and should have enough contrast from the background to be visualized by eye to ensure proper spot counting. We highly recommend hand-counting FISH spots with ImageJ's Cell Counter function when establishing a new pipeline for smFISH analysis.
6. After images are unbiasedly analyzed, select the image or crop of the image in ImageJ/FIJI such that it matches your quantifications (*see* **Note 11**).

4 Notes

1. TdT buffer is toxic and should be disposed of in a proper waste container.
2. Max absorption and extinction coefficient will depend on the fluorophore being used. For example, absorbance ATTO647N would be measured at 646 nm and an extinction coefficient of 1.5×10^5 M^{-1} cm^{-1}. We recommend checking the absorbance max and extinction coefficient on FPbase.
3. Adjust cell confluency as you would for other experiments. Overly confluent cells can cause cells to overlap with each other, reducing the single-cell power of smFISH.

4. A glass-bottom dish or coverslip (#1.5 H) is necessary for smFISH to ensure high-resolution images on 100×/1.4 NA oil objectives.
5. Other fixatives, such as methanol, can be used.
6. Using a cytoplasmic marker will be critical for automated downstream analysis of cytoplasmic RNAs. We routinely use the stress granule marker G3BP1 tagged with GFP to define the cytoplasm. Using a cytoplasmic marker will ensure proper segmentation of cells to garner single-cell information. Ensure the fluorophore used for your cytoplasmic marker does not spectrally overlap with your smFISH probes by checking FPbase.
7. RNase inhibitors are essential to add to all permeabilization or blocking steps to limit RNA degradation.
8. Similarly, to Note 6, a nuclear stain is necessary for downstream analysis for subcellular and single-cell localization of smFISH spots with BigFISH/FISH-Quant. We routinely use NucBlue with Hoechst #33342 for nuclei staining.
9. We typically image immediately or 1–3 days after final wash (with plates stored at 4°C). Longer-term storage can reduce fluorescence signal and image quality. This can be overcome by other preservation techniques.
10. We highly encourage researchers to use the BigFISH pipeline to analyze smFISH images. It encompasses segmentation, spot counting, cluster analysis, and subcellular single-cell localization features. Further, the documentation is clear and robust for performing analysis in Python.
11. We typically deconvolve our representative images first on Nikon Elements using a Richardson-Lucy Deconvolution algorithm.

Acknowledgments

We thank Moon lab members and this work was funded by a grant to S.L.M. from the NIGMS/NIH (GM146711). Writing – original draft: N.S.H.; Writing – review & editing: S.L.M.

References

1. Glock C, Heumüller M, Schuman EM (2017) mRNA transport & local translation in neurons. Curr Opin Neurobiol 45:169–177. https://doi.org/10.1016/j.conb.2017.05.005
2. Holt CE, Martin KC, Schuman EM (2019) Local translation in neurons: visualization and function. Nat Struct Mol Biol 26:557–566. https://doi.org/10.1038/s41594-019-0263-5
3. Alecki C, Rizwan J, Le P, Jacob-Tomas S, Comaduran MF, Verbrugghe M et al (2024) Localized molecular chaperone

synthesis maintains neuronal dendrite proteostasis. Nat Commun 15:10796. https://doi.org/10.1038/s41467-024-55055-7
4. de Queiroz BR, Laghrissi H, Rajeev S, Blot L, De Graeve F, Dehecq M et al (2025) Axonal RNA localization is essential for long-term memory. Nat Commun 16: 2560. https://doi.org/10.1038/s41467-025-57651-7
5. Trovisco V, Belaya K, Nashchekin D, Irion U, Sirinakis G, Butler R et al (2016) bicoid mRNA localises to the Drosophila oocyte anterior by random Dynein-mediated transport and anchoring. eLife 5:e17537. https://doi.org/10.7554/eLife.17537
6. Roth S, Neuman-Silberberg FS, Barcelo G, Schüpbach T (1995) cornichon and the EGF receptor signaling process are necessary for both anterior-posterior and dorsal-ventral pattern formation in Drosophila. Cell 81:967–978. https://doi.org/10.1016/0092-8674(95)90016-0
7. Clark I, Giniger E, Ruohola-Baker H, Jan LY, Jan YN (1994) Transient posterior localization of a kinesin fusion protein reflects anteroposterior polarity of the Drosophila oocyte. Curr Biol 4:289–300. https://doi.org/10.1016/s0960-9822(00)00068-3
8. Clark A, Meignin C, Davis I (2007) A Dynein-dependent shortcut rapidly delivers axis determination transcripts into the Drosophila oocyte. Development 134:1955–1965. https://doi.org/10.1242/dev.02832
9. Weil TT (2014) mRNA localization in the Drosophila germline. RNA Biol 11:1010–1018. https://doi.org/10.4161/rna.36097
10. Berleth T, Burri M, Thoma G, Bopp D, Richstein S, Frigerio G et al (1988) The role of localization of bicoid RNA in organizing the anterior pattern of the Drosophila embryo. EMBO J 7:1749–1756. https://doi.org/10.1002/j.1460-2075.1988.tb03004.x
11. Johnstone O, Lasko P (2001) Translational regulation and RNA localization in Drosophila oocytes and embryos. Annu Rev Genet 35:365–406. https://doi.org/10.1146/annurev.genet.35.102401.090756
12. Driever W, Nüsslein-Volhard C (1988) The bicoid protein determines position in the Drosophila embryo in a concentration-dependent manner. Cell 54:95–104. https://doi.org/10.1016/0092-8674(88)90183-3
13. Moon SL, Morisaki T, Khong A, Lyon K, Parker R, Stasevich TJ (2019) Multicolour single-molecule tracking of mRNA interactions with RNP granules. Nat Cell Biol 21:162–168. https://doi.org/10.1038/s41556-018-0263-4
14. Helton NS, Dodd B, Moon SL. (2025) Ribosome association inhibits stress-induced gene mRNA localization to stress granules. Genes Dev. 39(13–14):826–848. https://doi.org/10.1101/gad.352899.125
15. Wang ET, Taliaferro JM, Lee J-A, Sudhakaran IP, Rossoll W, Gross C et al (2016) Dysregulation of mRNA lcalization and translation in genetic disease. J Neurosci 36:11418–11426. https://doi.org/10.1523/JNEUROSCI.2352-16.2016
16. Engel KL, Arora A, Goering R, Lo H-YG, Taliaferro JM (2020) Mechanisms and consequences of subcellular RNA localization across diverse cell types. Traffic 21:404–418. https://doi.org/10.1111/tra.12730
17. Dahm R, Macchi P (2007) Human pathologies associated with defective RNA transport and localization in the nervous system. Biol Cell 99:649–661. https://doi.org/10.1042/BC20070045
18. Chin A, Lécuyer E (2017) RNA localization: Making its way to the center stage. Biochim Biophys Acta, Gen Subj 1861:2956–2970. https://doi.org/10.1016/j.bbagen.2017.06.011
19. Femino AM, Fay FS, Fogarty K, Singer RH (1998) Visualization of single RNA transcripts in situ. Science 280:585–590. https://doi.org/10.1126/science.280.5363.585
20. Raj A, van den Bogaard P, Rifkin SA, van Oudenaarden A, Tyagi S (2008) Imaging individual mRNA molecules using multiple singly labeled probes. Nat Methods 5:877–879. https://doi.org/10.1038/nmeth.1253
21. Imbert A, Ouyang W, Safieddine A, Coleno E, Zimmer C, Bertrand E et al. (2022) FISH-quant v2: a scalable and modular tool for smFISH image analysis. RNA 28:786–795. https://doi.org/10.1261/rna.079073.121
22. Gaspar I, Wippich F, Ephrussi A (2017) Enzymatic production of single-molecule FISH and RNA capture probes. RNA 23:1582–1591. https://doi.org/10.1261/rna.061184.117
23. Gaspar I, Wippich F, Ephrussi A (2018) Terminal deoxynucleotidyl transferase mediated production of labeled probes for single-molecule FISH or RNA capture. Bio Protoc 8:e2750. https://doi.org/10.21769/BioProtoc.2750

Chapter 8

Single-Molecule RNA Fluorescence In Situ Hybridization (smRNA-FISH) Labeling and Imaging of mRNA to Determine Subcellular Localization

Skyler Briggs and James M. Burke

Abstract

The protocol reported in this chapter describes methodologies for labeling of RNA by single-molecule RNA fluorescence in situ hybridization (smRNA-FISH), performing imaging confocal microscopy to image RNAs, and post-acquisition analyses. Specifically, we describe methods to design and generate smRNA-FISH probes, label RNAs in situ, and image RNAs. We also describe stepwise post-acquisition analyses using Image J to determine the absolute quantity, subcellular localization, and translation status of messenger RNAs. We provide examples of labeling full-length RNAs and their subspecies, quantifying RNA localization to organelles and ribonucleoprotein granules, and analyses of the translation status of individual mRNAs.

Keywords RNA imaging, smFISH, ImageJ analysis, smFISH quantification, Immunofluorescence, Confocal imaging, Subcellular localization, Single-cell analysis, RNA dynamics

1 Introduction

The field of RNA biology has progressed RNA from a pit stop between DNA and protein to appreciating its diverse functions in regulation, stability, and localization. Early methods such as radioactive in situ hybridization established the principle that transcripts can be visualized within cells [1]. The development of fluorescence in situ hybridization (FISH) and later RNA-FISH provided researchers with powerful tools to visualize transcripts within conserved cellular contexts [2, 3]. The development of single-molecule FISH (smFISH) revolutionized the field by enabling the detection of individual RNA molecules via multiple labeled probes [4, 5]. More recently, live-cell imaging systems and highly multiplexed approaches allow for real-time direct observation of RNA metabolism and transport in single cells [6].

James M. Burke is the lead contact.

Je-Hyun Yoon (ed.), *RNA Labeling*, Methods in Molecular Biology, vol. 3047,
https://doi.org/10.1007/978-1-0716-5352-4_8,

Many RNAs undergo dynamic regulation in response to stimuli, including increased expression, degradation, transport, or sequestration to specific regions of the cell [7]. In addition, these methods have revealed that these processes occur at specific subcellular compartments such as the nucleus, ribonucleoprotein (RNP) complexes, or are involved in cytosolic condensation events [8–10]. smFISH enables characterization of these processes and the location where these processes occur. These methods also provide robust ways to study responses to viral infection and define the mechanisms by which our immune system interfaces with viruses [11–14].

smFISH provides advantages over methods such as RNA-seq by providing single-cell data. This distinction is incredibly important for the following reason. If RNA-seq data indicate a transcript is upregulated during an antiviral response, does this mean each cell induced the upregulation of that gene or is a smaller percentage of cells in that population responsible for the increase? smFISH provides data that clarify this question. There are still advantages compared to single-cell RNA-seq where smFISH provides information on the location of the RNA during the experimental conditions. It may be the case that a transcript is not changing in number but is instead relocalized to function.

This chapter describes a robust protocol for RNA imaging that has been successfully applied to study RNA quantity and localization during broad experimental conditions. The described protocols provide a detailed explanation of immunofluorescence and RNA smFISH staining, imaging, and analysis. These tools enable robust measurement of RNA metabolism and localization during a wide set of experimental conditions. However, the described methods are not comprehensive, and there are many other ways to quantify and display data. We hope this guide provides a place to start and encourages creative new methods in the pursuit of understanding RNA biology.

2 Materials

2.1 Immunofluo rescence

- 10× nuclease-free PBS (dilute to 1× with nuclease-freewater).
- 75% ethanol.
- 12-mm glass coverslips (Fisher Scientific Co LLC: NC1418755).
- 12- or 24-well plate.
- Primary antibody.
- Secondary antibody.
- 32% paraformaldehyde (PFA).
- Vectashield with DAPI (Vector Laboratories: 101098-044).

2.2 Cell Culture

- 1× PBS.
- DMEM cell culture media.

- Trypsin.
- 15-mL tube.

2.3 Preparation of smFISH Probes

- TDT buffer (5×): 500-mM potassium cacodylate (pH 7.2), 10-mM $CoCl_2$, 1-mM DTT
- TDT enzyme
- Fluorescent-labeled dUTP (Jena Biosciences) in the desired fluorescent range
- Nuclease-free water
- 3-M sodium acetate
- Cold pure ethanol

2.4 smFISH

Hybridization buffer: 50% dextran sulfate, deionized formamide, 20× nuclease-free SSC, nuclease-free water, 25-mm syringe filter, 0.45-µm pore

20× nuclease-free SSC	1 mL
Deionized formamide	1 mL (*see* **Note 1**)
50% dextran sulfate	2 mL
Nuclease-free water	6 mL

Filter sterilize (0.45 µM), aliquot into 0.5 mL and store at −20°C.

GAPDH smFISH Probe

Buffer A: 20× nuclease-free SSC, nuclease-free water, deionized, Nalgene Rapid-Flow Sterile Disposable Filter Units, 500 mL, 0.45-µ mpore, 90-mm membrane

20× nuclease-free SSC	15 mL
Nuclease-free water	120 mL

Filter sterilize and store at 4°C.

Buffer B: 20× nuclease-free SSC, nuclease-free water, Nalgene Rapid-Flow Sterile Disposable Filter Units, 500 mL, 0.45-µm pore, 90-mm membrane

20× Nuclease-free SSC	15 mL
Nuclease-free water	135 mL

Filter sterilize and store at 4°C.

4% PFA (*see* **Note 2**)

32% PFA	10 mL
1× PBS	70 mL

Prepare aliquots of between 5 and 10 mL and store at −20°C until use.

- Polypropylene Petri dish
- Paper towel
- Deionized water
- Reagent reservoir
- Tweezers

2.5 Microscope Description

Nikon Eclipse Ti2 with a CFI60 Plan Apochromat Lambda D 100× Oil Immersion Objective Lens, N.A. 1.45, W.D. 0.13 mm, F.O.V. 25 mm, DIC, Spring Loaded. The filter set included:

C-FL DAPI Filter Set, High-Signal-Noise, Semrock Brightline®, Excitation: 356/30 nm (341–371 nm), Emission: 447/60 nm (417–477 nm), Dichroic Mirror: 409 nm

C-FL GFP/FITC/Cy2 Filter Set, High-Signal-Noise, Semrock Brightline®, Excitation: 466/40 nm (446–486 nm), Emission: 525/50 nm (500–550 nm), Dichroic Mirror: 495 nm

C-FL DSRed/TRITC/Cy3 Filter Set, High-Signal-Noise, Semrock Brightline®, Excitation: 554/23 nm (542.5–565.5 nm), Emission: 609/54 nm (582–636 nm), Dichroic Mirror:573 nm

C-FL Cy5 Filter Set, High-Signal-Noise, Semrock Brightline®, Excitation: 618/50 nm (593–643 nm), Emission: 698/70 nm (663–733 nm), Dichroic Mirror: 652 nm

2.6 Image Analysis

- Computer of minimum 4 GB RAM
- ImageJ, with "Batteries Included"

3 Methods

These methods detail a protocol for dual immunofluorescence assay and RNA smFISH following desired experimental conditions in confluent cultured cell layers. This allows for the visualization of proteins, subcellular structures, and RNAs in a cell. The following methods describe how to design oligos to target your RNA and subsequently order or couple those oligos to fluorophores for fluorescent imaging. Analysis techniques are described to characterize the subcellular localization of RNAs based on their colocalization with other structures and the quantification of RNAs based on their localization within the cell (Fig. 1).

3.1 Preparation of RNA smFISH Probes

To prepare RNA smFISH probes you have two options. The first is to order pre-prepared probes from a supplier such as Stellaris. The second option is to couple fluorescent probes to DNA oligos ordered from a supplier. The following includes instructions to order or prepare your own probes.

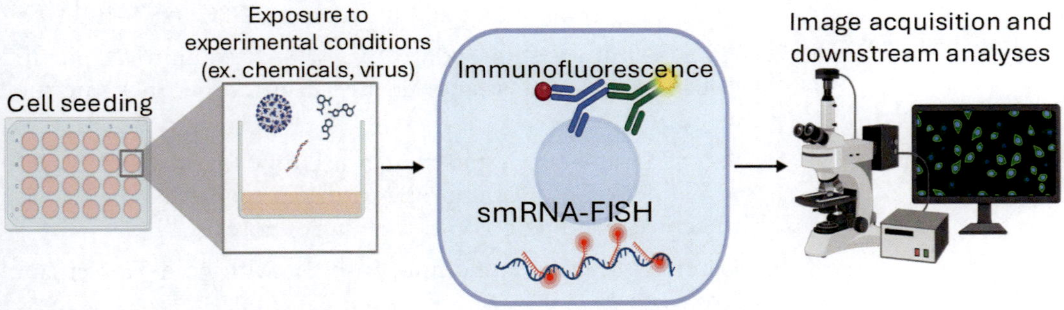

Fig. 1 Graphic overview of dual immunofluorescence and smFISH-RNA protocol where cells are seeded and exposed to experimental conditions. Cells exposed to experimental conditions are then probed via IF and/or smFISH to be imaged and analyzed via techniques described in the following sections

1. Define the RNA sequence(s) you would like to probe for and utilize smFISH probe design tool while following probe design recommendations. Stellaris maintains a public probe design tool that generates probe designs with high efficacy. Registration is free and the website includes resources on optimal probe design parameters* (*see* **Note 3**) (https://oligos.biosearchtech.com/products/rna-fish/stellaris-probe-designer).
2. Once a set of probes is designed, continue to either ordering pre-coupled probes or continue to Step 3 to couple probes with fluorophore yourself.
3. To couple the designed probes with fluorophore, order the probe set as DNA oligos from a supplier (IDT) at 200-μM concentration. Order in a 96-well plate format and label the plate for the RNA it contains probes for. Additionally, order the corresponding fluorophore to label the oligos with* (*see* **Note 4**).
4. Once probes and fluorophores are obtained, create a stock of mixed probes that will tile the entire RNA of interest. Pool 30 μL of each probe into a 1.5-mL tube for a mixed 200-μM stock. The total volume will depend on the number of probes ordered.
5. To prepare a fluorescently labeled probe, mix the following in a new 1.5-mL tube on ice, and add the enzyme last.

   - *Probe Reaction (on ice)*
     - 4-μL pooled probes from stock.
     - 10-μL TDT buffer.
     - 2-μL fluorescently labeled dUTP (Jena Biosciences).
     - 31-μL nuclease-free water.
     - 1.5-μL TDT enzyme (*see* **Note 5**).

6. Mix by gentle pipetting and incubate at 37°C for 24 h.
7. After 24 h incubation, add 50 µL of 3-M sodium acetate and 400 µL of ice-cold ethanol to the 1.5-mL tube. Incubate tube at −20°C for 24 h.
8. After 24 h at −20°C, spin the 1.5-mL tube for 30 min at 20,000 × *g* at 4°C. Ensure the probe forms an observable pellet with the corresponding color of the fluorophore.
9. Carefully aspirate liquid and wash 2× with cold 75% ethanol (*see* **Note 6**).
10. Resuspend pellet with 80 µL of nuclease-free water for the final RNA smFISH probe.

3.2 Coverslip Cell Seeding and Experimental Workflow

1. Place a 12- or 24-well plate, 12-mm glass coverslips, and ethanol in a cell culture biosafety cabinet. Place one coverslip in each well to be seeded for the experiment and add 0.5 mL 75% ethanol to each glass coverslip (ensure coverslip is submerged in ethanol and not floating on top).
2. Allow slips to be completely submerged in 75% ethanol for at least 2 min. Remove ethanol and place the plate lid at an angle to allow any remaining ethanol to evaporate. Leave the plate while cells are prepared for seeding (the method of cell detachment and seeding may need to be optimized for certain cell types; the following applies to most adherent immortal cell lines).
3. Bring cells into the biosafety cabinet. Aspirate media, wash with 2 mL of 1× PBS, and aspirate PBS. Add 1 mL 0.05% trypsin and place in incubator for 3–4 min until cells detach from flask.
4. Add 3 mL of pre-warmed media, move cells to a 15-mL conical tube, and spin at 500 × *g* for 5 min. Aspirate trypsin and media and resuspend in 1 mL of media. Count cells and seed cells at 80% confluency in the wells with coverslips. Evenly distribute cells using gentle cross-pattern motion and place back in 37°C incubator overnight for cells to adhere.
5. The following day, ensure cells are adhered to coverslips before continuing to the experimental conditions.
   *Both under- and over-confluent cell layers may cause experimental variability; therefore, it is important to use consistent cell confluency for consistent and reliable results.
6. Once cell confluency is optimal continue to experimental conditions. Common experimental conditions include exposure to compounds, viruses, and many other forms of cellular stress.
7. After completion of the experimental condition, aspirate media and wash with 1× PBS.
8. Fix the cells by pipetting 0.5 mL of 4% PFA to each well. Gently rock the plate and ensure slip is completely submerged in PFA.

Allow PFA to sit for 15 min at room temperature. Remove PFA and discard it in the waste bucket. *Paraformaldehyde is highly toxic and carcinogenic, carefully handle in a fume hood using protective gloves and a lab coat.

9. Wash the cells with 1 mL of 1× PBS and aspirate PBS. Permeabilize the cells by pipetting on 1 mL of 75% of ethanol. Ensure coverslips are completely submerged in ethanol. Let the cells permeabilize for at least an hour or let them sit overnight at 4°C.
   *Alternative permeabilization methods such as 0.1% Triton-X have produced robust results.
10. After cells have been fixed and permeabilized, proceed with smFISH and immunofluorescence staining.
    *Coverslips may be stored at 4°C for 3–5 days in 75% EtOH, while longer storage tends to produce worse results.

3.3 Immunofluorescence (see Note 7)

1. Take the 12-well plate out of the 4°C and aspirate off the 75% ethanol. Rinse the glass slips three times with nuclease-free 1× PBS, letting the cells sit in the third wash.
   *Cell layers may become unattached with aggressive washing. Add PBS to side of wells and rock plate for gentle washes.
2. In a 15-mL conical tube, dilute primary antibody to the pre-optimized concentration in nuclease-free 1× PBS. Aspirate PBS from each well.
3. Pipette 0.5 mL of diluted primary antibody to each well and ensure each coverslip is completely submerged in primary antibody solution.
4. Let primary antibody sit for 5 h at room temperature or overnight at 4°C.
5. Aspirate off the primary antibody and gently wash three times with nuclease-free 1× PBS letting the cells sit in the third wash.
6. In a 15-mL conical tube, dilute secondary antibody to the pre-optimized concentration in nuclease-free 1× PBS. Aspirate PBS from each well.
7. Pipette 0.5 mL of diluted secondary antibody to each well and ensure each coverslip is completely submerged in primary antibody solution.
8. Let secondary antibody sit for at least 2 h at room temperature while covered from light. Aspirate off the secondary antibody and rinse three times with nuclease-free 1× PBS.
9. Fix the cells by pipetting 0.5 mL of 4% PFA to each well. Gently rock to the plate and ensure slip is completely submerged in PFA. Allow PFA to sit for 15 min at room temperature. Remove PFA and discard in waste bucket. Rinse the slips three times with nuclease-free 1× PBS, letting the cells sit in the third wash.

3.4 smRNA-FISH

1. To prepare for smFISH, thaw hybridization buffer and GAPDH smFISH probe on ice.
2. Place Buffer A and deionized formamide in a fume hood and mix to obtain a 10% formamide by volume (e.g., 1-mL formamide + 9-mL buffer A). Leave Buffer A (10% formamide) solution until it reaches room temperature.
3. To prepare a hybridization chamber (Fig. 2), take a square polypropylene petri dish and cut a folded paper towel to fit flat to the bottom. Wet the paper towel with deionized water and place it in the petri dish. Take a piece of parafilm and cut to fit flush to the paper towel. Lay the parafilm on the paper towel and close the lid. Label the lid for the samples being dyed.
4. Ensure the probe and hybridization buffer are fully thawed before mixing. To a new 1.5-microcentrifuge tube, add hybridization buffer (50 μL/coverslip) and smFISH probe (0.5 μL/probe). Vortex briefly to suspend probes in hybridization buffer and place the tube on ice.

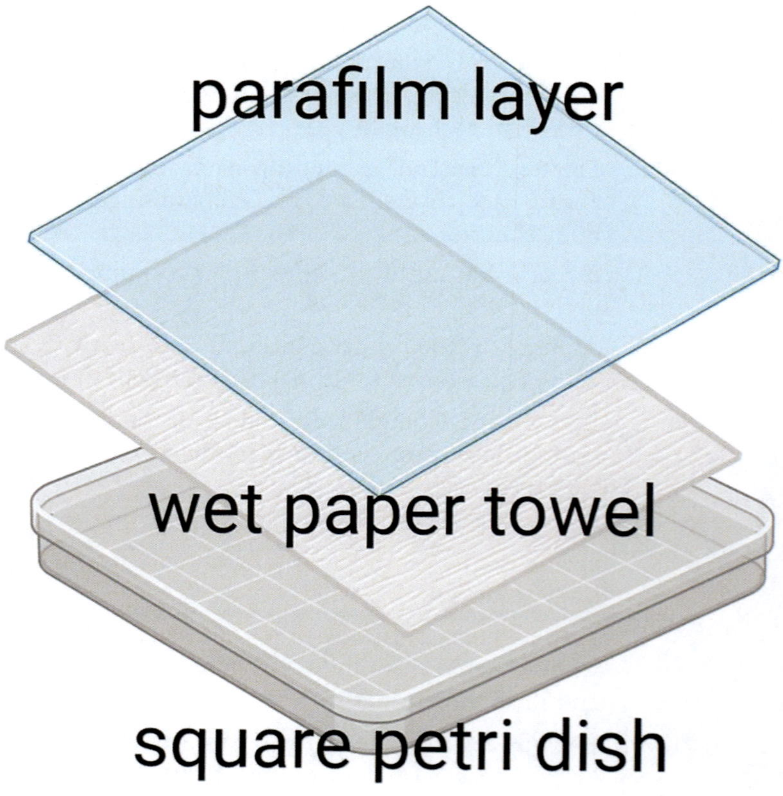

Fig. 2 Hybridization chamber for RNA smFISH. A parafilm layer on top of wet paper towel in a square petri dish forms a humid chamber for hybridization

5. Aspirate the PBS wash from the wells containing coverslips. Aliquot 0.5 mL of room-temperature Buffer A (10% formamide) solution to each well. Incubate coverslips in Buffer A for 5 min. *Formamide is toxic, carefully handle in fume hood using protective gloves and a lab coat.
6. While coverslips incubate in Buffer A (10% formamide) solution, place 50-μL aliquots of smFISH probe mixed in hybridization buffer on the parafilm in the hybridization chamber (Fig. 3). Place far enough away so that the drops will not interact with each other.
7. Carefully remove coverslips from the wells using tweezers and place them upside-down onto the hybridization buffer aliquots, the cell layer should be in contact with the smFISH probe + hybridization buffer mixture.
8. Place the lid back on the hybridization chamber and parafilm the chamber shut. Place the chamber in a 37°C incubator overnight.
9. Following incubation at 37°C, place Buffer A, Buffer B, and formamide in the fume hood. Prepare Buffer A with 10% formamide solution (1-mL formamide + 9-mL Buffer A). Aliquot enough Buffer B to add 0.5 mL to each coverslip and leave both aliquots at room temperature while returning stocks to 4°C.
10. Take the hybridization chamber out of the incubator and place it in the fume hood and use tweezers to carefully transfer the slips from the hybridization chamber and place them in the original orientation in the designated wells. The cell layer should now face up.
11. Aliquot 0.5 mL of Buffer A (10% formamide) to each well and let sit for 20 min. Ensure each coverslip is completely submerged in Buffer A (10% formamide).
12. After 20 min, aspirate off Buffer A and aliquot 0.5 mL of Buffer B to each well. Incubate at room temperature for 5 min. Organize and label glass microscope slides while incubating in Buffer B (include necessary information such as cell lines, time points, experimental conditions, and marker used in each channel).
13. Pipette 10 μL of Vectashield with DAPI for each coverslip. Carefully remove coverslips from wells with Buffer B and place on labeled microscope slides so cells are facing the Vectashield drops.
14. Aspirate off excess liquid around the slips and seal each slip with clear nail polish. Nail polish will impede imaging quality so attempt to coat edges and leave a significant coverslip area for imaging. Allow the nail polish to completely dry in a dark drawer before imaging.

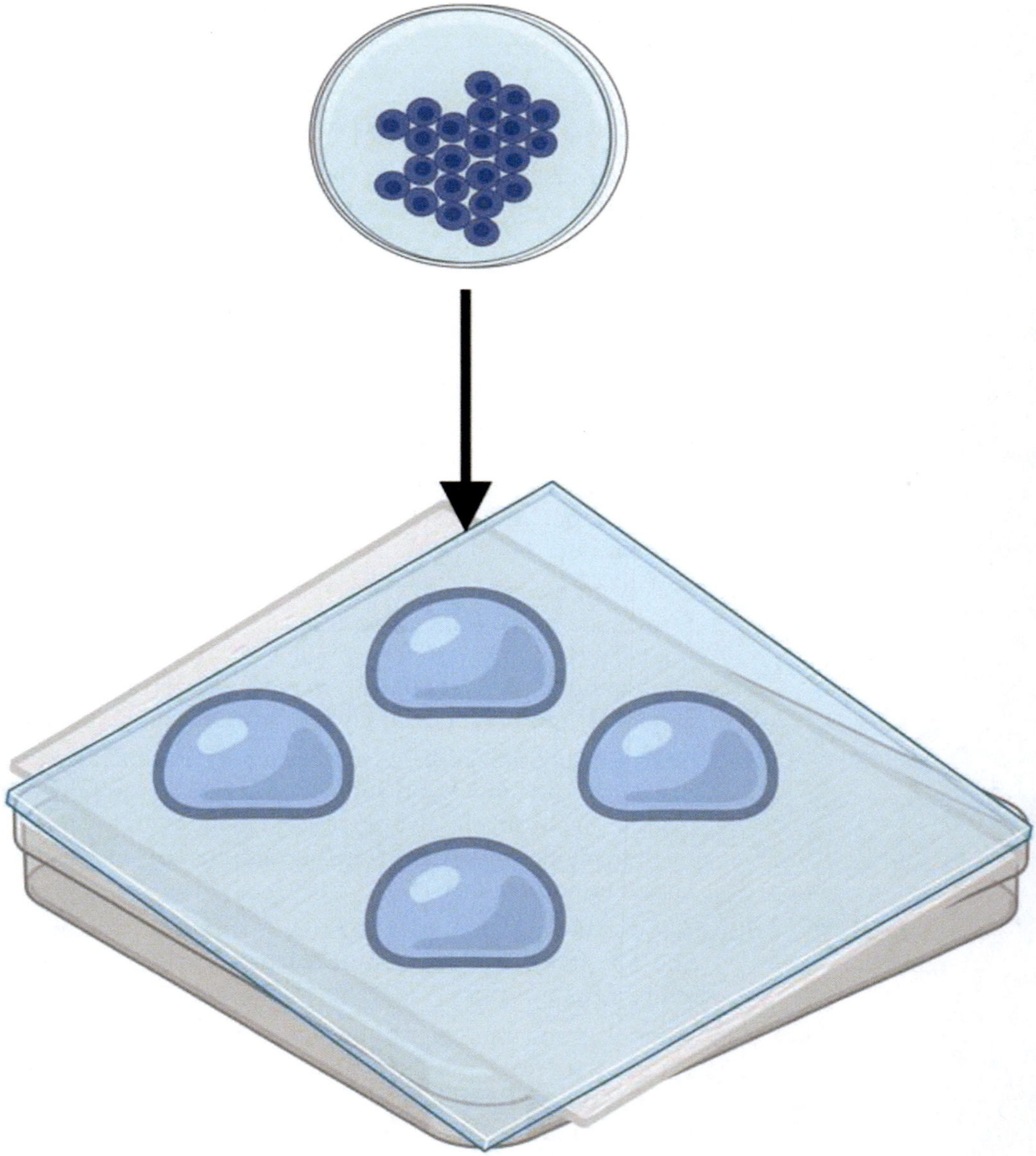

Fig. 3 Coverslips are placed cell-surface down on the drops of smFISH probe in hybridization buffer in hybridization chamber

3.5 Imaging Cells for Analysis

Imaging of cells will vary based on the microscope used and the RNA probes imaged. The demonstrated images used a Nikon Eclipse Ti2 with a 100×, 1.45 NA oil objective. Ensure that the exposure time and laser intensity are kept consistent throughout imaging.

1. Locate cells using the DAPI channel to eliminate bias toward active/inactive cells. Capture Z-stack images with each channel containing a marker using 2 μm z-spacing between 11 total slices.

2. Capture representative images of each condition ensuring enough are captured for quantification. It is recommended to capture images of 100 cells/condition minimally and to capture at varying magnification to ensure data captured are representative of cell population (*see* **Note 8**).

3.6 Image Analysis

Before starting: The software used to analyze images is ImageJ or FIJI, we recommend downloading Fiji: ImageJ with "Batteries Included" which comes prepackaged with many useful plugins. This program is free and can be downloaded here: https://fiji.sc.

There are repositories of ImageJ macros available for free download, and we have also uploaded a folder of macros to GitHub for free download and use, https://github.com/skylerbriggs/smFISH-Quantification-Macros.

The smFISH method allows for the quantification of the number of RNA within single cells. The following instructions detail the strategy to quantify RNA numbers and methods to increase throughput of quantification. In genes with high transcripts number, individual RNA may not be discerned (e.g., rRNA or peak viral RNA loads). However, cytoplasmic intensity correlates with RNA content [15].

The quantification of the number of RNAs within single cells provides excellent data on increased RNA expression or degradation. The imaging of RNA in situ additionally allows for the determination of RNA localization within single cells. The following details methods used to determine where RNA is located within the cell and how to measure and demonstrate location/colocalization based on co-staining with other markers.

3.7 Quantification of Absolute mRNA Copies per Cell

1. The first step of analysis is to format the collected images from the file type used by the microscope software to tiff format. The image file type will change depending on the brand of microscope used but ImageJ opens and converts most file formats without a problem.
2. To convert the image, drag and drop the image into the ImageJ taskbar. This will likely open the image using the Bio-Formats plugin already included in the "Batteries Included" version of ImageJ. To optimally open the image, ensure "Hyperstack" is selected in stack viewing and the "Use Virtual Stack" box is checked to minimize memory used for the image. Nothing else should be checked and press "OK."
3. There are many tools available, but the primary tools are under the "Image" menu. Click "Image>Color>Channels Tool" to open a box where each channel captured during imaging can be selected or deselected. Click "Image>Adjust>Brightness/Contrast" to adjust the view of each channel (*see* **Note 9**).

4. Once image channels are adjusted, convert the image to tiff. Select "File>Save As>Tiff…" and choose a location to save the image. Complete this process for each image to be analyzed (*see* **Note 10**).
5. To quantify the number of smFISH RNA puncta within a cell, first open the tiff file of the image to be analyzed. Select "Image>Adjust>Threshold…" to threshold the image to only include smFISH puncta. Adjust upper and lower bound intensity values, select "Dark background" and "Don't reset range" then select "Apply" (Fig. 4A) (*see* **Note 11**).
6. To measure the number of RNA puncta in a single cell, use the "Freehand Tool" to draw an outline around the cell. Select "Analyze>Analyze Particles…", change Size: "0–2," Circularity: "0–1," Show: "Nothing," and select "Summarize"

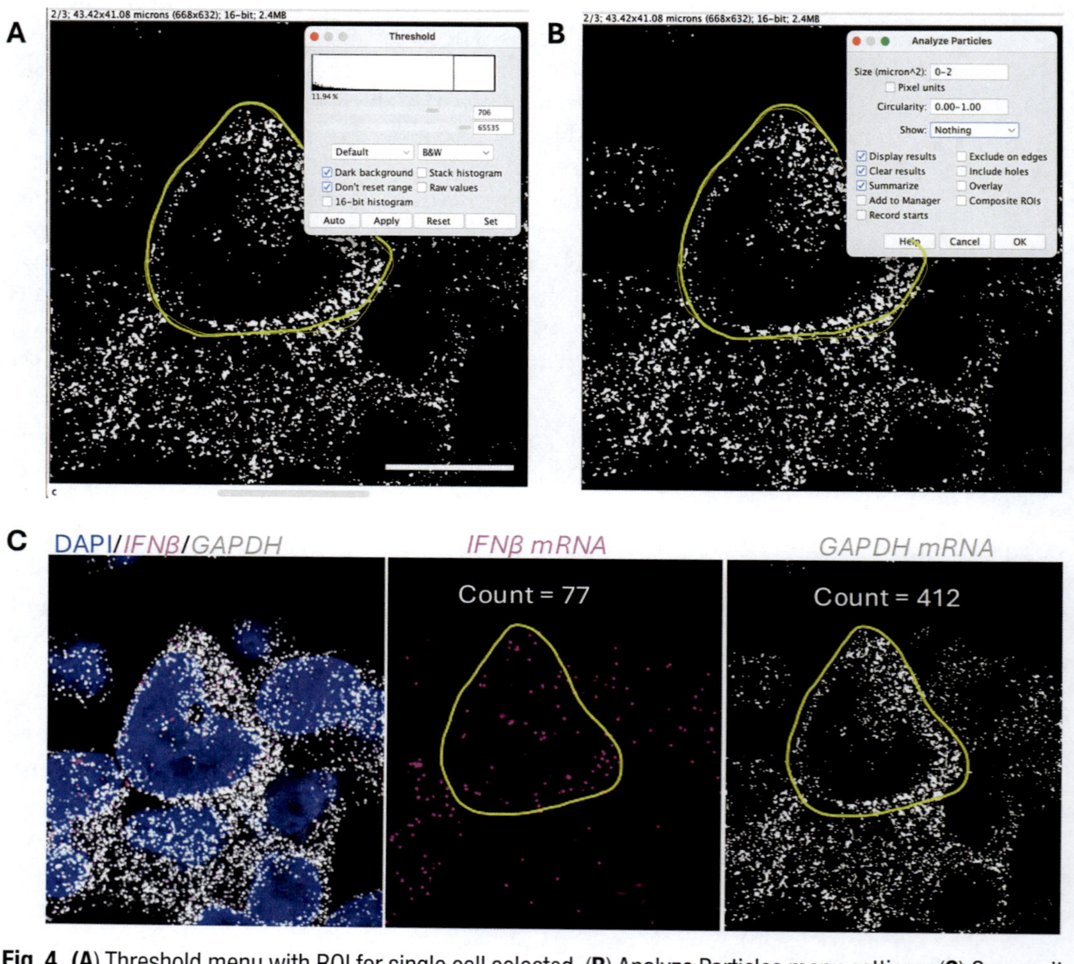

Fig. 4 (**A**) Threshold menu with ROI for single cell selected. (**B**) Analyze Particles menu settings. (**C**) Composite image of cell analyzed for particle count and respective smFISH dots within ROI

(Fig. 4B). Select "OK" to output the measured number of particles in the region of interest selected.

7. Repeat this process for each channel. Copy and paste the data to an Excel sheet for analysis and repeat this step for each cell.
8. To increase analysis throughput, utilize the ImageJ macro available for free download at the GitHub link provided above. The macro follows the same steps outlined above and allows for quantification of all cells in a field of view at once (Fig. 5A, B). The folder includes instructions for this increased throughput method.

3.8 Line Trace

The following quantification is performed on cells infected with West Nile Virus (WNV) and probed for OASL (FITC channel), 5′ WNV RNA (TRITC channel), and WNV subgenomic flavivirus RNA (sfRNA) (Cy5 channel). When cells are infected with WNV they can activate RNase L and degrade viral RNA. However, portions of the viral genome are not degraded but are shuttled to RNP complexes. smFISH was performed on cells expressing OASL-GFP to determine if OASL granules and sfRNA in RNP complexes are colocalized.

1. To perform a line trace, open the image to be analyzed. Use the "Straight Line" tool to draw a line through a region of interest (Fig. 6A).

   *Use an appropriate scale for the measurement to be made and choose a representative cell.
2. Select "Analyze>Plot Profile" and select "Live" for live data on the channel selected. Use the "Channels" tool to select

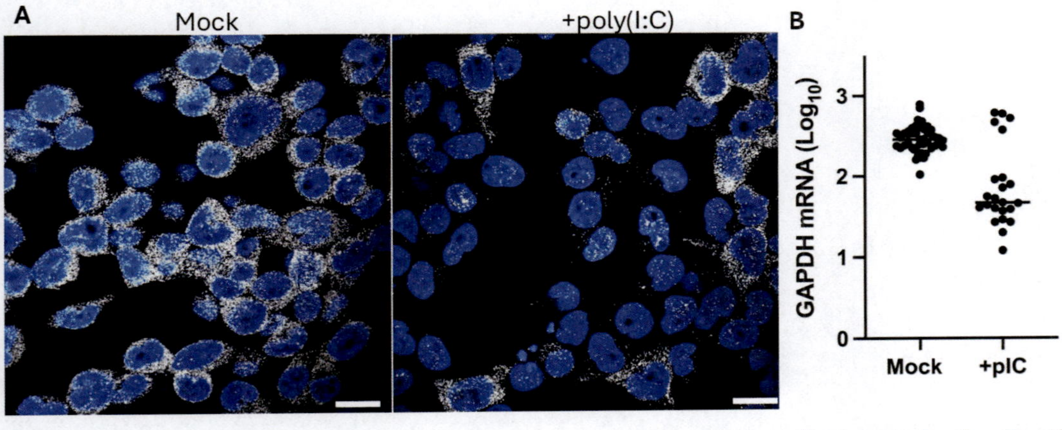

Fig. 5 (**A**) smFISH for *GAPDHmRNA* in mock and poly(I:C) treatment conditions. Treatment of cells with pIC activates RNase L, resulting in the degradation of cellular mRNA. (**B**) Quantification of smFISH dots for cells in (**A**) using high-throughput macros

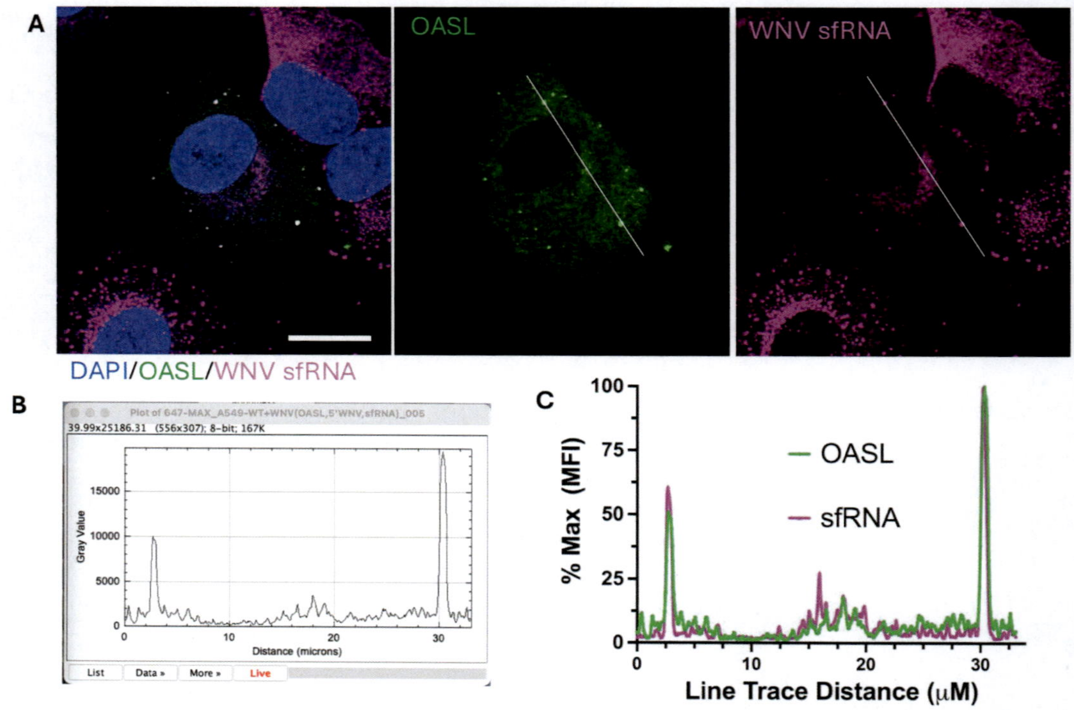

Fig. 6 (**A**) Representative image of a cell infected with West Nile Virus then probed for OASL and 5′ WNV RNA with a line trace through the region of interest. (**B**) Plot profile live menu for OASL intensity. (**C**) Line-trace plot of the % maximum MFI of each channel for line in (**A**)

the channel to be measured. Ensure the plot updates for each channel (Fig. 6B).

3. To save the measurement for each channel, select "Data>Copy All Data" and paste to an Excel sheet. Alternatively, the data can be saved to a spreadsheet.
4. Utilize the spreadsheet to display intensity of each channel along the line. In many cases, it is best to represent the data as a percent of maximum (Fig. 6C).

3.9 Cytoplasmic vs Nuclear RNA

Upon stimulation, cells will often produce new RNA or degrade RNA, a process known as RNA metabolism. In addition, those RNA will localize to specific regions of cells which provide information on the biology of that RNA. To demonstrate the quantification of RNA in cellular locations, cells were treated with IFNβ and subsequently probed via smFISH for new *IFNβ mRNA* transcripts. The following method is used to quantify RNA within cellular regions.

1. To quantify the number of smFISH RNA puncta within a subcellular region, first open the tiff file of the image to be analyzed.

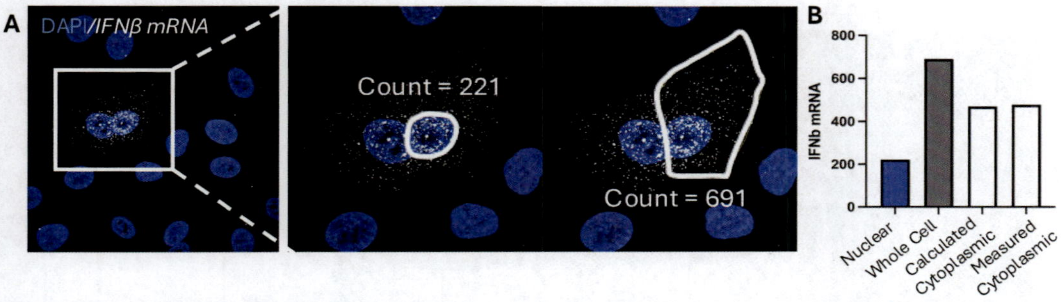

Fig. 7 (**A**) Cells treated with IFNB and probed for IFNB mRNA via smFISH. Inset shows the transcript count for the circled regions. (**B**) Transcript number quantified from (**A**)

2. Use the "Channels" tool to select the channel with the smFISH data to be quantified.
3. Select "Image>Adjust>Threshold..." to threshold the image to only include smFISH puncta. Adjust upper and lower bound intensity values, select "Dark background" and "Don't reset range" then select "Apply."
4. To measure the number of RNA puncta in a region, use the "Freehand Tool" to draw an outline around the area of interest. In this case, circle the nucleus.
5. Select "Analyze>Analyze Particles...," change Size: "0–2," Circularity: "0–1," Show: "Nothing," and select "Summarize." Select "OK" to output the measured number of particles in the region of interest outlined (Fig. 7A).
6. Repeat Steps 2 and 3 but circle the entire cell with the "Freehand Tool" to quantify whole cell transcript number and subtract nuclear transcript number for cytoplasmic number. Alternatively, circle only the cytoplasm (Fig. 7B).

3.10 Localization of RNAs to RNP Complexes

Upon stimulation, cells may undergo liquid-liquid phase separation and completely reorganize. Protein and RNA reorganization can be observed and measured using immunofluorescence and smFISH. In the following example, cells were treated with sodium arsenite to form stress granule RNP complexes. The cells were then probed via smFISH for *AHNAK mRNA* The following method is used to quantify the colocalization of cytosolic RNA and RNP complexes during stress response.

1. To quantify the colocalization of RNP complexes and RNA, open the tiff file of the image to be quantified.
2. Select "Image>Color>Split Channels"
3. Select "Analyze>Colocalization>Coloc 2" to open colocalization analysis window.

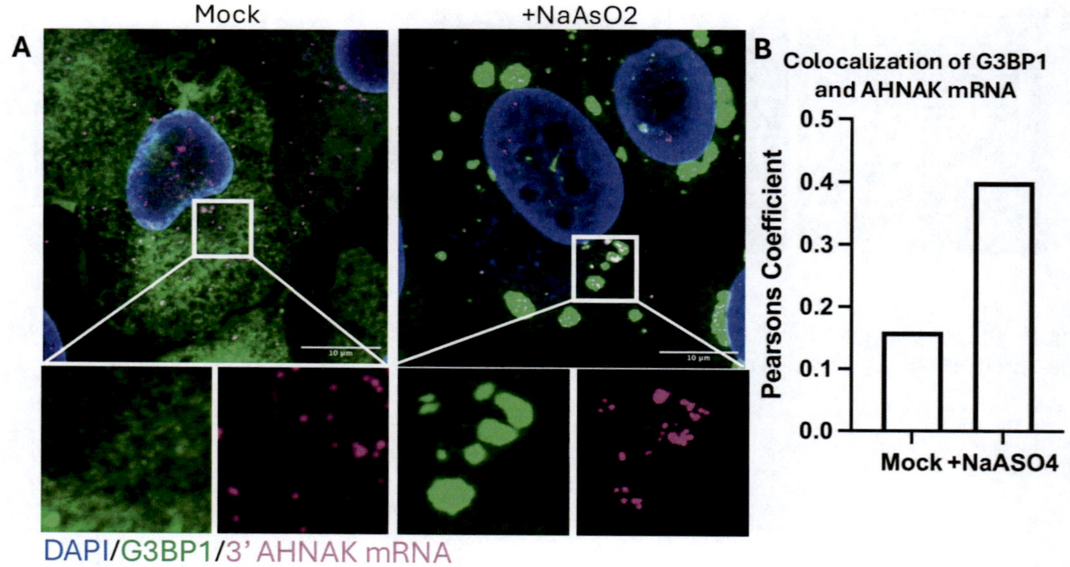

Fig. 8 (**A**) Cells exposed to sodium arsenite form stress granules. (**B**) Sequestration of mRNA to stress granules quantified as correlation coefficient

4. Select the two channels to be measured for colocalization, this example uses channel 2 (GFP) and channel 4 (AHNAK mRNA) (Fig. 8A). Use Costs threshold regression and select the boxes for setting and algorithms (*see* **Note 12**).
5. Select "OK" and allow for quantification algorithm to complete. Save the PDF result to a desired analysis location and copy the data to a spreadsheet. This example uses Pearson's coefficient (Fig. 8B) to quantify colocalization, while the specific correlation may be tuned for unique experiments.

3.11 Differential Localization of RNAs and RNA Subspecies

There are many examples in which the localization or quantification of specific RNA isoforms or RNA subspecies is of interest. To do this, probes can be targeted to distinct regions of an RNA and differentially labeled. For example, flaviviruses have an 11-kb RNA genome and generate subgenomic RNAs, known as sfRNAs. To differentiate sfRNAs from full-length genomes, probes targeting unique the 5′-end labeled with ATT0-550 and 3′-end labeled with ATT-633 were generated as described in [10] (Fig. 9A). Thus, the full-length genome RNA foci contain both the 5′-end (green) and 3′-end (magenta), and thus appear white (Fig. 9B). However, the sfRNAs stains with the 3′-end probe but lack the 5′-end (green) and thus are magenta foci. Similarly, subgenomic SARS-CoV-2 RNAs were determined to differentially localize in the cell in comparison to the full-length genome [16]. In principle, this assay could be used to identify specific splice isoforms and RNA processing intermediates

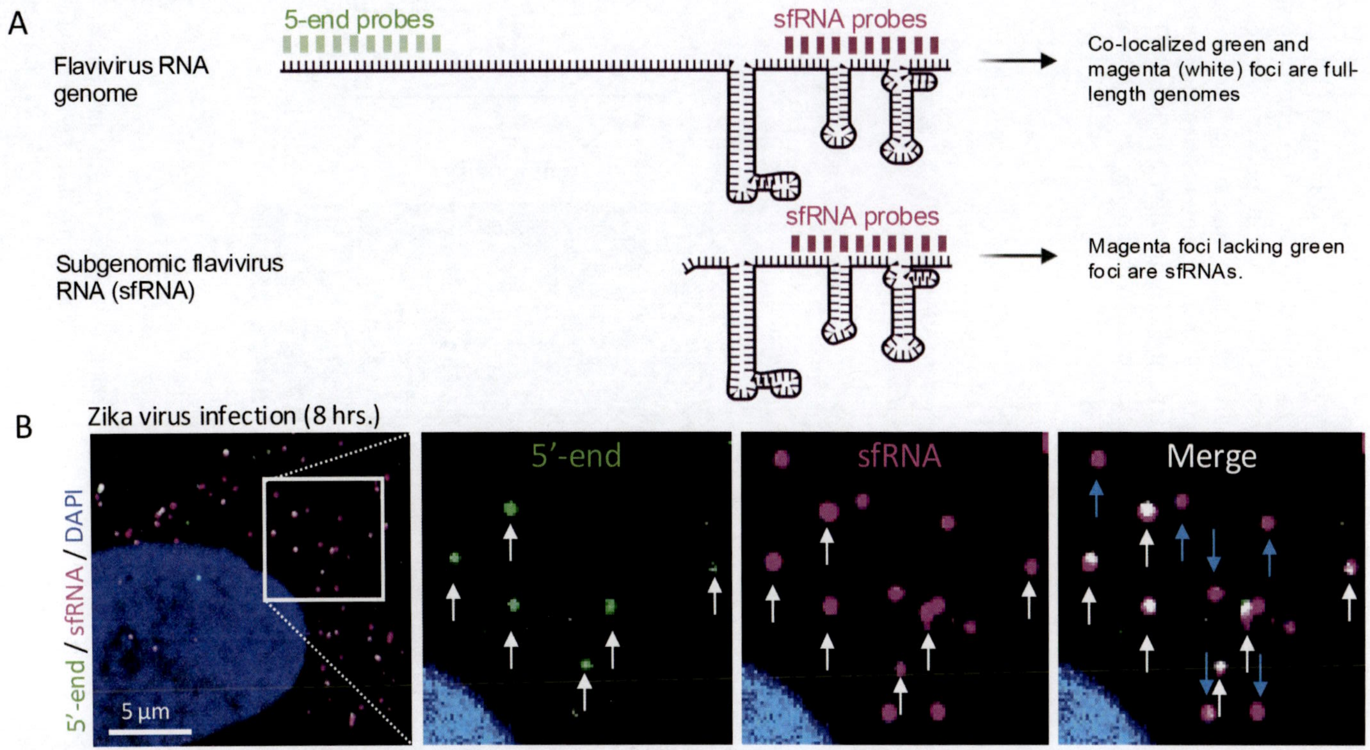

Fig. 9 (**A**) Graphical representation of smRNA-FISH labeling strategy. (**B**) Images of smRNA-FISH labeling described in (A) where 5′-end is degraded and only sfRNA remains 8 h after infection with Zika virus

as long as a sufficient length of RNA (~200 nt) is available to target for hybridization of probes.

3.12 Single mRNA Translation Status

smRNA-FISH enables the quantification of translating RNA by measuring the distance between 5′ and 3′ ends [17]. Upon ribosome binding, RNA ends are pulled further apart and correspondingly upon translation inhibition RNA ends are pulled together (Fig. 10A). This distance can be measured to determine the translation state of the RNA, which provided surprising results that redefined how we think about the state of RNA during translation. The following method is used to measure RNA end distance to determine mRNA translation status.

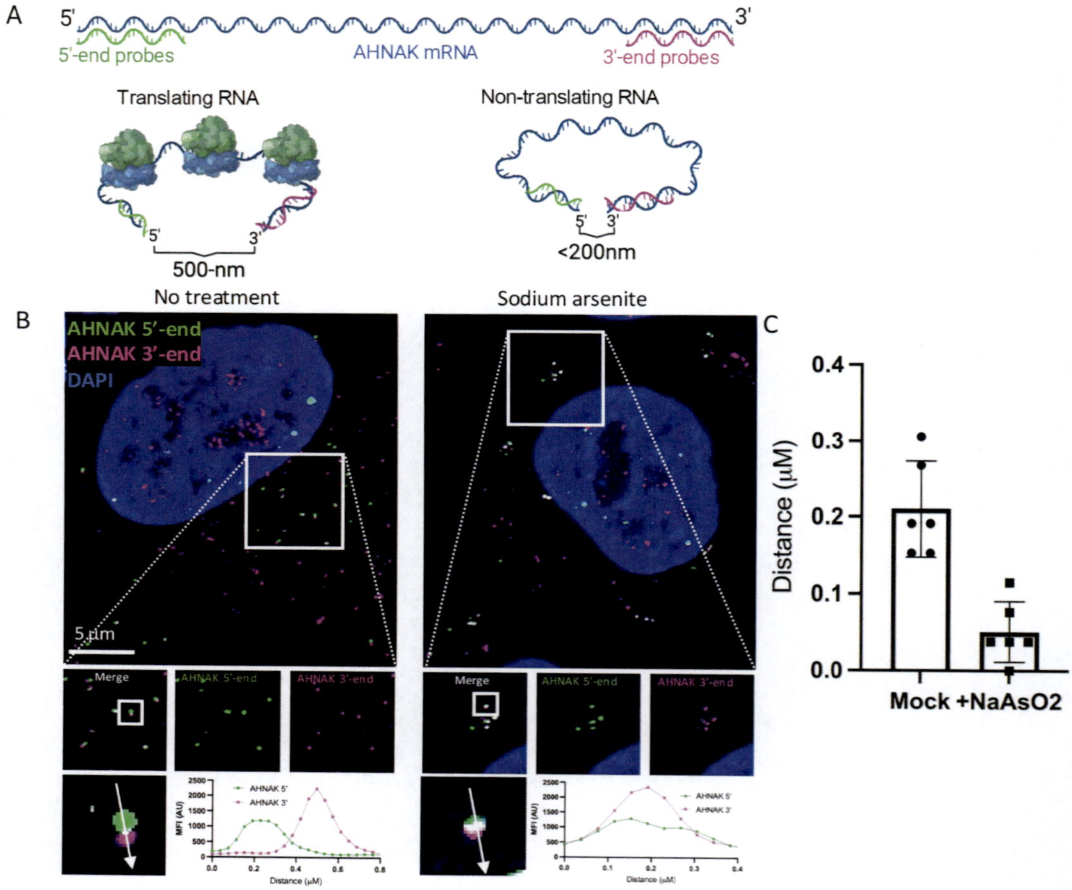

Fig. 10 (**A**) Graphical representation of mRNA probe labeling strategy to measure the translation state. (**B**) A549 cells treated with or without sodium arsenite, which inhibits translation. Cells treated with sodium arsenite exhibit reduced distance in 5′ and 3′ probes as demonstrated by imaging and line-trace plots. (**C**) Quantification of the distance between the maximal intensity of the 5′- and 3′-end trace plots, as represented in (**B**)

1. Open the image containing smRNA-FISH images with uniquely fluorescent end probes. It is best to use a single z-slice for this assay to select for RNA parallel to the z-plane.
2. Identify two puncta close together indicating they are labeling the same mRNA. Draw a line using the Straight-Line tool through the puncta of interest.
3. Select "Analyze>Plot Profile" and select "Live" for live data on the channel selected. Use the "Channels" tool to select the channel to be measured. Ensure the plot updates for each channel.
4. Save the intensity plot for the desired channels and save them to a spreadsheet for further analysis. Ensure to use the same line for each channel for proper measurement.
5. In the spreadsheet, find the distance between the maximum values of intensity for each channel. This difference is the distance between 5′ and 3′ ends of the RNA. These intensity plots can be visualized using the same methods as 1. Line trace above (Figs. 6 and 10B) and the difference in distance is represented as a simple bar plot (Fig. 10C).

4 Notes

1. Formamide is toxic, carefully handle it in a fume hood using protective gloves and a lab coat.
2. Note: Paraformaldehyde is highly toxic and carcinogenic, carefully handle in a fume hood using protective gloves and a lab coat.
3. Minimally, design no less than 25 oligos per probe set for each RNA. This may require altering the nucleotide size of each probe; do not use less than 18 nt for each probe.
4. Ensure the fluorophores are compatible with the microscope to be utilized and do not overlap with other fluorescent markers to be used in the same experiment. Multiple probes can be used in the same experiment and therefore must be coupled to unique fluorophores.
5. TdT enzyme is highly sensitive to temperature, keep it at −20°C ice box.
6. Do not resuspend the pellet, carefully pipette and aspirate ethanol without disturbing pellet.
7. If not performing immunofluorescence in combination with smFISH, proceed to RNA smFISH (following section).

8. Additional imaging parameters may be necessary for each experiment. Consult the manufacturer of the equipment for specific questions on imaging capabilities.
9. Adjust each channels brightness and contrast to view the signal optimally. This is a subjective decision but adjusting will not change the raw data held in the image.
10. z-stack images may be stacked for a maximum projection image. To save as a maximum projection select "Image>Stacks>Z Project…" and select the range of z-slices to stack. This z-stack image is saved with the same process. Ensure you use the number of slices needed for the data. If quantifying the total number of smFISH puncta, use a z-stack, where if quantifying nuclear vs cytoplasmic, a single z-slice may be more representative as it demonstrates a single slice through a cell.
11. Record the value used to threshold and use the same values for each cell analyzed across the experiment for consistent measurement.
12. A detailed description of the function of each algorithm can be found on ImageJ's website.

References

1. Gall JG, Pardue ML (1969) Formation and detection of RNA-DNA hybrid molecules in cytological preparations. Proc Natl Acad Sci USA 63:378–383. https://doi.org/10.1073/pnas.63.2.378
2. Bauman JGJ, Wiegant J, Borst P, van Duijn P (1980) A new method for fluorescence microscopical localization of specific DNA sequences by in situ hybridization of fluorochrome-labelled RNA. Exp Cell Res 128:485–490. https://doi.org/10.1016/0014-4827(80)90087-7
3. Singer RH, Ward DC (1982) Actin gene expression visualized in chicken muscle tissue culture by using in situ hybridization with a biotinated nucleotide analog. Proc Natl Acad Sci USA 79:7331–7335. https://doi.org/10.1073/pnas.79.23.7331
4. Femino AM, Fay FS, Fogarty K, Singer RH (1998) Visualization of single RNA transcripts in situ. Science 280:585–590. https://doi.org/10.1126/science.280.5363.585
5. Raj A, van den Bogaard P, Rifkin SA et al (2008) Imaging individual mRNA molecules using multiple singly labeled probes. Nat Methods 5:877–879. https://doi.org/10.1038/nmeth.1253
6. Xia C, Colognori D, Jiang XS, et al (2025) Single-molecule live-cell RNA imaging with CRISPR–Csm. Nat Biotechnol 43:2023–2030. https://doi.org/10.1038/s41587-024-02540-5
7. Buxbaum AR, Haimovich G, Singer RH (2015) In the right place at the right time: visualizing and understanding mRNA localization. Nat Rev Mol Cell Biol 16:95–109. https://doi.org/10.1038/nrm3918
8. Briggs S, Blomqvist EK, Cuellar A, Correa D, Burke JM (2025) Condensation of human OAS proteins initiates diverse antiviral activities in response to West Nile virus. Genes Dev 39(19-20):1162–1181. https://doi.org/10.1101/gad.352725.125 PMID: 40588419; PMCID: PMC12487701
9. Cusic R, Burke JM (2024) Condensation of RNase L promotes its rapid activation in response to viral infection in mammalian cells. Sci Signal 17:eadi9844. https://doi.org/10.1126/scisignal.adi9844
10. Watkins JM, Burke JM (2024) RNase L-induced bodies sequester subgenomic flavivirus RNAs to promote viral RNA decay. Cell Rep 43:114694. https://doi.org/10.1016/j.celrep.2024.114694
11. Burke JM, Lester ET, Tauber D, Parker R (2020) RNase L promotes the formation of unique ribonucleoprotein granules distinct from stress granules. J Biol Chem 295:1426–1438. https://doi.org/10.1074/jbc.RA119.011638
12. Burke JM, Ripin N, Ferretti MB et al (2022) RNase L activation in the cytoplasm induces

aberrant processing of mRNAs in the nucleus. PLoS Pathog 18:e1010930. https://doi.org/10.1371/journal.ppat.1010930
13. Burke JM, Gilchrist AR, Sawyer SL, Parker R (2021) RNase L limits host and viral protein synthesis via inhibition of mRNA export. Sci Adv 7, 23. https://doi.org/10.1126/sciadv.abh2479
14. Burke JM, Moon SL, Matheny T, Parker R (2019) RNase L Reprograms translation by widespread mRNA turnover escaped by antiviral mRNAs. Mol Cell 75:1203–1217.e5. https://doi.org/10.1016/j.molcel.2019.07.029
15. Little SC, Tikhonov M, Gregor T (2013) Precise developmental gene expression arises from globally stochastic transcriptional activity. Cell 154:789–800. https://doi.org/10.1016/j.cell.2013.07.025
16. Burke JM, St Clair LA, Perera R, Parker R (2021) SARS-CoV-2 infection triggers widespread host mRNA decay leading to an mRNA export block. RNA 27:1318–1329. https://doi.org/10.1261/rna.078923.121
17. Khong A, Parker R (2018) mRNP architecture in translating and stress conditions reveals an ordered pathway of mRNP compaction. J Cell Biol 217:4124–4140. https://doi.org/10.1083/jcb.201806183

Check for updates

Chapter 9

RNA Labeling of Nucleolar Transcripts by 5-Ethynyl Uridine Click Chemistry

Juyeon Lee and Kyung-Ha Lee

Abstract

The nucleolus is a dynamic membraneless organelle where ribosomal RNA (rRNA) transcription, processing, and ribonucleoprotein assembly occur. Visualizing nascent RNA within nucleolar condensates reveals how RNA metabolism and phase separation reorganize under stress. This chapter provides a practical protocol to label nascent RNA using 5-ethynyl uridine (5-EU), followed by copper-catalyzed azide–alkyne cycloaddition for fluorescent detection. The method is straightforward, works across mammalian cells, and robustly reports nucleolar transcription under basal and stress conditions (heat shock; low-dose actinomycin D). Step-by-step instructions, reagent setup, image analysis metrics, troubleshooting, safety, and expected results are included so that researchers can reproducibly implement the assay.

Keywords Nucleolus, RNA labeling, 5-ethynyl uridine, Click chemistry, Nascent RNA, Nucleolar stress, Actinomycin D, Heat shock, Condensates

1 Introduction

The nucleolus is the largest nuclear condensate and the site of rRNA synthesis and ribosome biogenesis. Its multiphase architecture (fibrillar center, dense fibrillar component, granular component) reorganizes rapidly when RNA polymerase I activity fluctuates or when cells experience heat, oxidative, or transcriptional stress [1, 2]. Such perturbations can cause nucleolar fragmentation and the formation of nucleolar caps at the periphery of the nucleolus, which represent re-localized nucleolar proteins under transcriptional stress, or the release of RNA–protein aggregates into the nucleoplasm—hallmarks that are highly informative for condensate biology and stress signaling [3–6].

To monitor RNA synthesis with sufficient temporal control, metabolic labeling with modified ribonucleosides is widely used [7]. We focus on a single, high-impact method—5-Ethynyl uridine (5-EU) metabolic labeling + click detection—for nucleolus-centric studies

Je-Hyun Yoon (ed.), *RNA Labeling*, Methods in Molecular Biology, vol. 3047,
https://doi.org/10.1007/978-1-0716-5352-4_9,

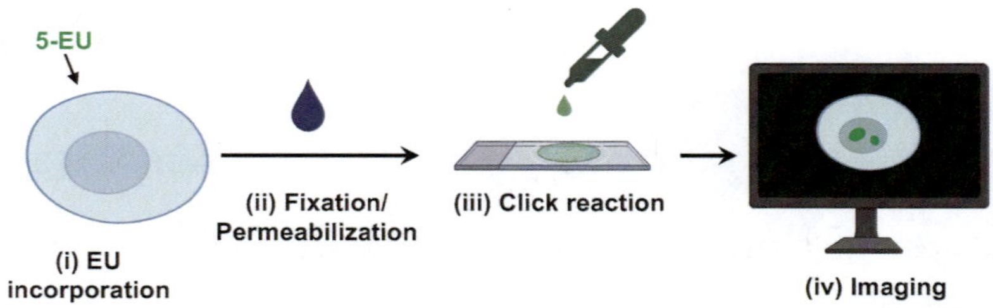

Fig. 1 Workflow of 5-EU nucleolar labeling. (i) EU incorporation; (ii) fixation/permeabilization; (iii) click reaction ($CuSO_4$–THPTA–ascorbate order; azide dye); (iv) imaging and ROI quantification

(Fig. 1). Use cases include mapping nucleolar transcription under stress, pulse–chase–style recovery measurements, and quantifying nucleolar vs. nucleoplasmic labeling as a condensate readout [8, 9]. In particular, low-dose actinomycin D (≈0.05 μg/mL) selectively inhibits RNA polymerase I–dependent rRNA transcription, leading to the formation of nucleolar caps at the periphery as a hallmark of nucleolar stress, while leaving RNA polymerase II largely unaffected [10–12]. For completeness, we note that aptamer- and immuno-based RNA labeling methods also exist; however, they are not covered as primary protocols in this chapter.

5-EU is incorporated into nascent RNA by cellular polymerases and then clicked to an azide fluorophore, yielding a bright, specific signal without antibodies [7, 13]. Because rRNA transcription is intense, nucleoli provide a strong internal positive signal after short EU pulses [8]. Protocols dedicated to EU labeling and EU-based chasing emphasize the practical details that govern click efficiency and reproducibility; this chapter aligns with those conventions (e.g., order and ratio of copper catalyst, THPTA ligand, and sodium ascorbate).

2 Materials

2.1 Cell Culture and General Reagents

1. HeLa or U2OS cells: Cell lines with prominent nucleoli; standard growth.
2. DMEM, high glucose.
3. Fetal bovine serum (FBS), heat-inactivated.
4. Penicillin–Streptomycin.
5. Glass coverslips.
6. 1× PBS.

2.2 Labeling and Fixation Reagents

1. 200 mM, 5-Ethynyl uridine (5-EU) in DMSO or water; –20°C.
2. 4% Paraformaldehyde (PFA) in PBS; fresh.
3. 0.5% Triton X-100 in PBS (permeabilization).

2.3 Click Chemistry Components

1. 50 mM Copper(II) sulfate ($CuSO_4$) in water.
2. 50 mM THPTA ligand in water.
3. 100 mM Sodium ascorbate
4. 5 mM Azide fluorophore (Alexa Fluor 488/594/647 azide) in DMSO.

Critical: The order and timing of additions ($CuSO_4$ → THPTA → sodium ascorbate → sample) and the Cu:THPTA ratio are decisive for click efficiency.

2.4 Counterstains and Optional Reagents

1. 1 μg/μL DAPI in PBS.
2. Anti-fibrillarin antibody.
3. Anti-nucleolin antibody.
4. Secondary antibodies.
5. Mounting medium (antifade).

2.5 Equipment

1. CO_2 incubator (37°C, 5% CO_2).
2. Pipettes.
3. Tips.
4. Humidified staining chamber.
5. Fluorescence/confocal microscope.
6. ImageJ/Fiji.

3 Methods

An overview of the experimental workflow and timing is summarized in Table 1.

3.1 Cell Preparation and Stress Induction

1. Plate HeLa or U2OS cells on coverslips to reach 50%–70% confluency on the day of labeling.
2. Optional stresses:
   - Heat shock: 42 °C, 30 min.
   - Actinomycin D (low dose): 0.05 μg/mL, 1 h for Pol I selective inhibition and nucleolar cap formation [14]. At this concentration, actinomycin D primarily targets RNA polymerase I, which accounts for its utility in nucleolar stress assays.
3. Include untreated controls for baseline nucleolar activity.

3.2 5-EU Pulse Labeling

1. Prepare pre-warmed medium with 0.2–1 mM 5-EU
2. Replace culture medium with EU medium; incubate 15–60 min (37°C). Short pulses (~10–15 min) already illuminate nucleoli;

Table 1
Timeline summary of 5-EU nucleolar RNA labeling workflow

Timeline summary	
Step	**Duration**
Cell plating	24 h before labeling
Stress induction (heat shock or actinomycin D)	30–60 min
5-EU pulse labeling	15–60 min
Fixation	10 min
Permeabilization	5 min
Click chemistry reaction	30 min
Counterstaining & antibody staining	1–2 h
Imaging	Variable

Overview of the sequential steps required for 5-EU metabolic labeling, click chemistry detection, and imaging analysis in mammalian cells. Estimated durations are provided to assist experimental planning

longer pulses increase the signal-to-noise ratio, making nucleolar labeling clearer and easier to quantify.

3.3 Fixation and Permeabilization

1. Rinse once with PBS; fix with 4% PFA, 10 min, RT.
2. Wash 3× PBS; permeabilize 0.5% Triton X-100, 5 min; wash 3× PBS.
3. Optional quench: 50 mM NH_4Cl, 5 min to lower background.

3.4 Click Chemistry Detection (Copper-Catalyzed Azide-Alkyne Cycloaddition)

1. Prepare fresh per coverslip (~500 μL total):
   - 20 μL 50 mM $CuSO_4$ (2 mM final)
   - 20 μL 50 mM THPTA (2 mM final)
   - 20 μL 100 mM sodium ascorbate (4 mM final; add last)
   - 2 μL 5 mM azide-fluorophore (~20 μM final)
   - Balance with PBS/HEPES to 500 μL.
2. Mix quickly; a brownish color indicates Cu(I) formation. Incubate 30 min, dark, RT; wash 3× for 5 min in PBS + 1% BSA or 0.1% Tween-20. THPTA stabilizes Cu(I) and protects fluorophores/proteins [9, 15, 16].

3.5 Counterstaining and Immunostaining (Optional)

1. DAPI 1 μg/mL, 5 min.
2. Nucleolar markers (e.g., fibrillarin 1:500, 1 h) → wash → secondary (1:1000, 30 min) → wash → mount.
3. Order: click → immunostaining (to minimize the potential effects of copper and reducing agents).

3.6 Imaging and Basic Quantification

1. Acquire images on a confocal or high-numerical-aperture epifluorescence microscope system with fixed exposure and gain across conditions.
2. Untreated cells show bright nucleolar EU signal and diffuse nucleoplasmic labeling; actinomycin D suppresses nucleolar EU and may reveal nucleolar caps at the periphery. Heat shock effects are typically reversible.

3.7 Controls and Experimental Design

1. Negative controls: (i) No EU; (ii) No ascorbate in click mix; (iii) No $CuSO_4$. All define background and click dependency.
2. Positive controls: EU-labeled cells without stress treatment (no heat shock, no actinomycin D), representing Pol I–active nuclei with strong nucleolar EU incorporation.
3. Stress controls: Include heat shock and low-dose actinomycin D to benchmark reversible vs. persistent nucleolar reorganization.
4. EU pulse selection: For rRNA-centric reads, start at 0.2–0.5 mM, 15–30 min; extend to 30–60 min for whole-nucleus comparisons.
5. Pulse–chase (optional): EU 15–30 min → chase with EU-free medium 15–120 min; track recovery/turnover.
6. Biological replicates: ≥3 independent cultures; ≥3 imaging fields per replicate.
7. Acquisition order: Always image control first to avoid parameter drift.

3.8 Quantitative Image Analysis

1. ROI definition
   - Nuclear mask: threshold DAPI to define the nucleus.
   - Nucleolar mask: from fibrillarin/nucleolin channel (preferred) or phase-contrast morphology with intensity-inversion/thresholding.

2. Primary metrics
   - Nucleolar fraction = Σ(EU_intensity_nucleolus)/ Σ(EU_intensity_nucleus)
   - Nucleolar contrast = mean(EU_nucleolus)/ mean (EU_nucleoplasm)

3. Workflow (ImageJ/Fiji)
   - (1) Load images → (2) Create nuclear mask (DAPI) → (3) Create nucleolar mask (fibrillarin or morphology) → (4) Measure mean/sum EU intensities per ROI → (5) Export CSV → (6) Statistics (*t*-test or Mann–Whitney; report effect size and CI) [17, 18].

4. Acquisition standardization: identical laser power, detector gain, exposure, and z-sampling across groups.

4 Notes and Troubleshooting

4.1 General Notes

1. Order matters in click chemistry; add sodium ascorbate last. THPTA must be present before reduction to stabilize Cu(I) [9, 15, 19, 20].
2. For RNA capture instead of imaging, substitute biotin-azide and follow the pull-down steps.
3. If DMSO stocks are used for EU/azide, include DMSO vehicle controls.
4. HeLa/U2OS are recommended for optimization; extend to other human/animal lines after baseline titration.

4.2 Troubleshooting Table

Common technical issues and suggested solutions are summarized in Table 2.

Table 2
Troubleshooting guide for 5-EU nucleolar RNA labeling and click detection

Problem	Possible cause	Solution
Weak nucleolar EU signal	EU dose or pulse too short	Increase to 0.5–1 mM EU or 30–60 min pulse
High background	Excess unreacted dye;	Wash 3× for 5 min with PBS + 1% BSA/0.1% Tween20;
	Washes insufficient	Reduce dye to ~20 μM
Poor click efficiency	Wrong addition order;	Add THPTA before sodium ascorbate;
	Aged ascorbate	Use fresh sodium ascorbate;
		Check Cu:THPTA
Loss of nucleolar structure	Over-fixation or harsh permeabilization	Limit PFA to 10 min;
		Consider saponin instead of Triton for fragile lines
Photobleaching	Excess illumination	Lower laser power;
		Shorter exposure;
		Antifade mounting
Replicate variability	Cell density/stress inconsistency	Plate to 50%–70%;
		Time/temperature-controlled stress application

Common technical issues encountered during 5-EU incorporation, copper-catalyzed azide–alkyne cycloaddition, and fluorescence imaging, together with likely causes and recommended solutions to improve signal intensity, reduce background, and preserve nucleolar morphology

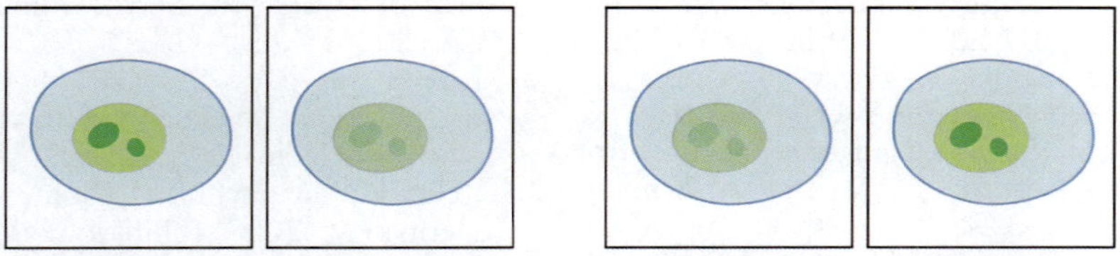

Fig. 2 Schematic representation of labeling outcomes. Control cells show strong nucleolar EU; actinomycin D reduces nucleolar signal; heat-shock samples demonstrate reversible recovery in nucleolar fraction over time

5 Expected Results

1. Control: Bright nucleolar EU with diffuse nucleoplasmic labeling; nucleoli are the highest intensity compartments after short pulses (Fig. 2).
2. Heat shock (42°C, 30 min): Transient reduction and re-equilibration of nucleolar EU upon recovery; suitable for reversibility assays with pulse–chase.
3. Actinomycin D (0.05 μg/mL, 1 h): Marked drop in nucleolar fraction; nucleolar cap patterns emerge at the periphery when co-stained with fibrillarin/nucleolin.
4. Quantitative ranges (typical): Nucleolar fraction decreases by ~40%–70% under Pol I inhibition relative to untreated, cell-line, and conditions dependent.
5. Reproducibility: CV across biological replicates ≤20% with standardized acquisition and ROI pipelines.

Acknowledgments

This work was supported by the Bio & Medical Technology Development Program of the National Research Foundation of Korea (NRF) grant funded by the Ministry of Science & ICT (RS-2024-00399681), an NRF grant funded by the Korean government (MSIT) (RS-2025-16070351).

References

1. Lafontaine DLJ, Riback JA, Bascetin R, Brangwynne CP (2021) The nucleolus as a multiphase liquid condensate. Nat Rev Mol Cell Biol 22(3):165–182
2. Boulon S, Westman BJ, Hutten S, Boisvert FM, Lamond AI (2010) The nucleolus under stress. Mol Cell 40(2):216–227
3. Yang K, Yang J, Yi J (2018) Nucleolar stress: hallmarks, sensing mechanism and diseases. Cell Stress 2(6):125–140
4. Lawrimore J, Kolbin D, Stanton J, Khan M, de Larminat SC, Lawrimore C et al. (2021) The rDNA is biomolecular condensate formed by polymer-polymer phase separation and is

sequestered in the nucleolus by transcription and R-loops. Nucleic Acids Res 49(8):4586–4598
5. Frottin F, Schueder F, Tiwary S, Gupta R, Korner R, Schlichthaerle T et al (2019) The nucleolus functions as a phase-separated protein quality control compartment. Science 365(6451):342–347
6. Nollen EA, Salomons FA, Brunsting JF, van der Want JJ, Sibon OC, Kampinga HH (2001) Dynamic changes in the localization of thermally unfolded nuclear proteins associated with chaperone-dependent protection. Proc Natl Acad Sci USA 98(21):12038–12043
7. Jao CY, Salic A (2008) Exploring RNA transcription and turnover *in vivo* by using click chemistry. Proc Natl Acad Sci USA 105(41):15779–15784
8. Dvorackova M, Fajkus J (2018) Visualization of the nucleolus using ethynyl uridine. Front Plant Sci 9:177
9. Gupta H, Lee JR, Hoffman KB, Min KW, Yoon JH (2025) RNA decay assay: 5-ethynyl-uridine labeling and chasing. Methods Mol Biol 2863:139–149
10. Cortes CL, Veiga SR, Almacellas E, Hernandez-Losa J, Ferreres JC, Kozma SC et al (2016) Effect of low doses of actinomycin D on neuroblastoma cell lines. Mol Cancer 15:1
11. Andersen JS, Lam YW, Leung AK, Ong SE, Lyon CE, Lamond AI et al (2005) Nucleolar proteome dynamics. Nature 433(7021):77–83
12. Andersen JS, Lyon CE, Fox AH, Leung AK, Lam YW, Steen H et al (2002) Directed proteomic analysis of the human nucleolus. Curr Biol 12(1):1–11
13. Paredes E, Das SR (2011) Click chemistry for rapid labeling and ligation of RNA. ChemBioChem 12(1):125–131
14. Kalita K, Makonchuk D, Gomes C, Zheng JJ, Hetman M (2008) Inhibition of nucleolar transcription as a trigger for neuronal apoptosis. J Neurochem 105(6):2286–2299
15. Fantoni NZ, El-Sagheer AH, Brown T (2021) A Hitchhiker's guide to click-chemistry with nucleic acids. Chem Rev 121(12):7122–7154
16. Hong V, Presolski SI, Ma C, Finn MG (2009) Analysis and optimization of copper-catalyzed azide-alkyne cycloaddition for bioconjugation. Angew Chem Int Ed Engl 48(52):9879–9883
17. Schneider CA, Rasband WS, Eliceiri KW (2012) NIH image to ImageJ: 25 years of image analysis. Nat Methods 9(7):671–675
18. Schindelin J, Arganda-Carreras I, Frise E, Kaynig V, Longair M, Pietzsch T et al (2012) Fiji: an open-source platform for biological-image analysis. Nat Methods 9(7):676–682
19. Presolski SI, Hong VP, Finn MG (2011) Copper-catalyzed azide-alkyne click chemistry for bioconjugation. Curr Protoc Chem Biol 3(4):153–162
20. Martin S, Kim CY, Coller J (2025) Assessment of mRNA decay and calculation of codon occurrence to mRNA stability correlation coefficients after 5-EU metabolic labeling. Methods Mol Biol 2863:151–182

Chapter 10

Simultaneous Visualization of NEAT1_2 and Paraspeckle Proteins by Sequential RNA-FISH and Immunofluorescence

Joonwoo Lee and Youngdae Gwon

Abstract

Paraspeckles are membrane-less subnuclear assemblies whose organization depends on the long isoform of nuclear paraspeckle assembly transcript 1 (NEAT1_2), which scaffolds paraspeckle proteins into a core–shell architecture. Disruption of the layered structure within paraspeckles occurs when paraspeckle components are depleted, potentially leading to functional impairment. Here, we describe a protocol for simultaneous visualization of NEAT1_2 middle region and transactive response DNA-binding protein 43 (TDP-43), which are enriched in paraspeckle core and shell, respectively, in U2OS human osteosarcoma cells. Sequential RNA-fluorescence in situ hybridization (RNA-FISH) and immunofluorescence, combined with high-resolution confocal microscopy, enable delineation of the paraspeckle core–shell structure. This method provides a robust approach for probing paraspeckle remodeling across diverse biological conditions.

Keywords Paraspeckle, Core–shell structure, NEAT1, NONO, TDP-43, RNA-FISH, IF

1 Introduction

Paraspeckles are a type of ribonucleoprotein (RNP) nuclear bodies, assembled from the long non-coding RNA nuclear paraspeckle assembly transcript 1 (NEAT1) together with paraspeckle proteins (PSPs) [1]. Of the two NEAT1 isoforms transcribed from the same promoter, the short isoform NEAT1_1 (3.7 kb) and the long isoform NEAT1_2 (22 kb) in humans, NEAT1_2 provides the architectural scaffold for paraspeckle formation and structural integrity by accommodating PSPs [2–4].

Paraspeckles are typically observed in punctate or elongated forms, where the latter becomes more prominent when NEAT1_2 transcription is elevated [5, 6]. The elongated paraspeckles are built upon the assemblies of NEAT1_2 RNPs, which act as a copolymer block of the micellization process [7]. Super resolution microscopy has revealed that paraspeckles adopt a unique core–shell structure [8]. NEAT1_2 folds into a loop in which the central region

Je-Hyun Yoon (ed.), *RNA Labeling*, Methods in Molecular Biology, vol. 3047,
https://doi.org/10.1007/978-1-0716-5352-4_10,

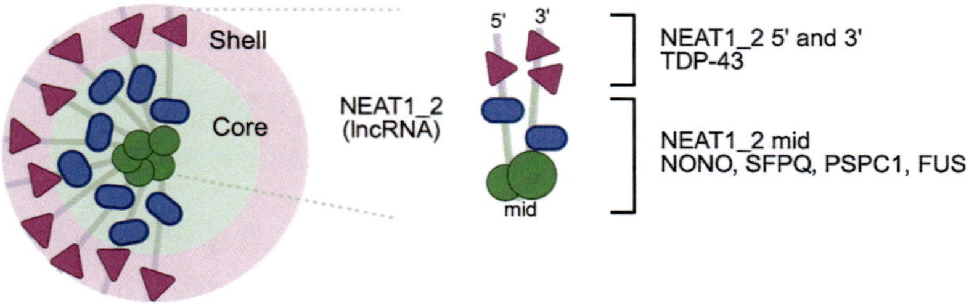

Fig. 1 Illustration of the core–shell structure of paraspeckles. The architectural lncRNA NEAT1_2 provides a scaffold, with its 5′ and 3′ ends localized to the shell and its middle region oriented toward the core. Paraspeckle proteins such as NONO, SFPQ, PSPC1, and FUS occupy the core, while TDP-43 is found at the shell region

accommodates PSPs such as non-POU domain-containing octamer binding protein (NONO), splicing factor proline- and glutamine-rich (SFPQ), paraspeckle component 1 (PSPC1), and fused in sarcoma (FUS), thereby forming the inner core. In contrast, the flanking 5′ and 3′ ends of NEAT1_2 localize to the outer shell with other groups of proteins, including transactive response DNA-binding protein 43 (TDP-43) (Fig. 1).

Recent studies have shown that deletion of specific NEAT1 domains and depletion of paraspeckle shell proteins disrupt the core–shell architecture of paraspeckles and their compartmentalization from adjacent nuclear bodies [4, 9, 10]. Given that NEAT1_2 and associated PSPs contribute to gene regulation by sequestering microRNAs and retaining transcripts containing inverted Alu (IRAlu) sequence, it is important to examine not only changes in paraspeckle abundance but also fine-scale structural remodeling across diverse biological contexts [11–15].

Here, we describe a method to capture paraspeckle architecture by simultaneously visualizing NEAT1_2 central region and TDP-43 which are, respectively, resident in core and shell of paraspeckles using RNA-FISH with domain-targeted probes in combination with immunofluorescent (IF).

2 Materials

Equivalent reagents, consumables, or equipment from other suppliers may be used if they provide comparable conditions to those described below. All reagents used were molecular-grade, and those designated as RNase-free were additionally treated using the method described later in this protocol. All reagents were stored under the manufacturer's recommended conditions.

2.1 Generation of U2OS Stably Expressing NONO-GFP Cell Line

U2OS stably expressing NONO-GFP cells are hereafter referred to as U2OS NONO-GFP stable cells.

1. The human osteosarcoma cell line, U2OS.
2. pEGFP-N1-NONO construct (*see* **Note 1**).
3. Transfection reagent, Viafect™ (#E498A, Promega).
4. Reduced serum medium, Opti-Mem™ (#31985-062, Gibco).
5. Selection antibiotics, G-418 (#ant-gn-1, InvivoGen).
6. Polystyrene round-bottom tube.
7. FACSAria™ Fusion flow cytometers (BD Bioscience).
8. Complete medium: Dulbecco's Modified Eagle Medium (DMEM) with 10% fetal bovine serum (FBS), 100 U/ml penicillin, and 1% Glutamax™ (#35050-061, Gibco).
9. Six-well cell culture plate and 60-mm cell culture dish.

2.2 Preparation of Poly-D-Lysine-Coated Coverslips

1. 0.1 mg/ml poly-D-lysine hydrobromide in DW.
2. 18-mm round coverslip.
3. 12-well cell culture plate.
4. 1× Phosphate-buffered saline (PBS): 137 mM NaCl, 2.7 mM KCl, 10 mM Na_2HPO_4, and 1.8 mM KH_2PO_4.

2.3 Fixation and Permeabilization for Sequential RNA-FISH and IF

1. U2OS NONO-GFP stable cell (see Sect. 2.1).
2. Poly-D-lysine-coated 18-mm round coverslips in a 12-well culture plate (see Sect. 2.2).
3. RNase-free 1× PBS: 1× PBS using RNase-free distilled water (*see* **Note 2**).
4. 4% Paraformaldehyde (PFA) solution in PBS (#J19943-1L, Thermo Scientific).
5. Permeabilization solution: 0.2% Triton™ X-100 in RNase-free 1× PBS.

2.4 Hybridization of RNA-FISH Probe for Staining NEAT1_2

1. Stellaris® FISH Probes, Human NEAT1 Middle Segment with Quasar® 570 Dye (#SMF-2037-1, LGC Biosearch Technologies) (*see* **Note 3**).
2. 10% Hybridization buffer: Add 900 µl of Stellaris® RNA FISH Hybridization Buffer (#SMF-HB1-10, LGC Biosearch Technologies) into 100 µl deionized formamide (#F9037-100ML, Merck).
3. Humidified chamber: A 100-mm culture plate dish was prepared by evenly lining the bottom with a flat, DW-saturated paper towel, on top of which a single layer of parafilm was placed to prevent evaporation of the probe solution from beneath the coverslip.

4. Wash buffer A: Add 2 ml of Stellaris® RNA FISH Wash Buffer A (#SMF-WA1-60, LGC Biosearch Technologies) into 8 ml of RNase-free 1× PBS.
5. Wash buffer B: Add 880 ml of RNase-free water to Stellaris® RNA FISH Wash Buffer B (#SMF-WB1-20, LGC Biosearch Technologies).
6. Fine-tipped forceps.

2.5 Sequential IF for Staining TDP-43

1. RNase-free 1× PBS.
2. Blocking solution: 1% bovine serum albumin (BSA) in RNase-free 1× PBS.
3. TDP-43 polyclonal antibody (#10782-2-AP, Proteintech).
4. Goat anti-Rabbit IgG (H+L) Highly Cross-Adsorbed Secondary Antibody, Alexa Fluor™ 647 (#A-21245, Invitrogen).
5. Microscope slides.
6. ProLong™ Gold Antifade Mountant (#P36934, Invitrogen).
7. Fine-tipped forceps.

2.6 Confocal Microscopy

High-resolution LSM980 confocal laser scanning microscope (Zeiss) (*see* **Note 4**).

3 Methods

3.1 Generation of Cell Line Stably Expressing NONO-GFP

1. The day before transfection, seed 2.0×10^5 U2OS cells per six-well cell culture plate in 1.5 ml of complete DMEM.
2. Incubate the cells at 37°C and 5% CO_2 in a cell incubator.
3. After 18–20 h of seeding, mix 100 μl of Opti-Mem™ with 4.5 μl of Viafect™ in polystyrene tube, then add 1.5 μg of pEGFP-N1-NONO plasmid (DNA:Viafect ratio = 1:3, μg:μl) for each well.
4. After gentle mixing, incubate at room temperature for 10 min, then add the mixture to the seeded cells.
5. At 24 h post-transfection, confirm transfection efficiency using a fluorescence microscope, then replace the transfection medium with complete DMEM containing 1 mg/ml G418 (selection medium).
6. Replace the medium with fresh selection medium every 3 days for 2weeks.
7. Sort NONO-GFP-positive U2OS cells using a FACSAria™ Fusion flow cytometers.

The U2OS NONO-GFP stable cell line is maintained in 60-mm culture dishes containing complete DMEM at 37°C in a humidified

5% CO_2 incubator, and subcultured by trypsinization when confluence reaches ~80%.

3.2 Preparation of Poly-D-Lysine-Coated Coverslips

1. Place an 18-mm coverslip at the bottom of a well in a 12-well plate.
2. Drop 500 μl of 0.1 mg/ml poly-D-Lysine solution onto the coverslip.
3. Incubate at room temperature for 1 h.
4. Rinse the coverslip with 1× PBS three times.
5. Air-dry the coated coverslip at room temperature for at least 1 h.

3.3 Fixation and Permeabilization for Sequential RNA-FISH and IF

All incubation steps should be performed under light-protected conditions.

The following protocol is based on adherent cells seeded on poly-D-lysine-coated 18-mm round coverslip placed in a 12-well cell culture plate. Volumes can be adjusted according to the preferred system.

The schematic workflow of sequential RNA-FISH and IF is illustrated in Fig. 2.

1. Grow U2OS NONO-GFP cells on poly-D-lysine-coated 18-mm round coverslip in a 12-well culture plate.
2. Aspirate the culture medium and wash with 1 ml of RNase-free 1× PBS.
3. Fix with 1 ml of 4% PFA at room temperature for 10 min (*see* **Note 5**).
4. Wash with 1 ml of RNase-free 1× PBS three times at room temperature for 5 min.
5. Incubate with permeabilization solution at room temperature for 5 min (*see* **Note 6**).
6. Wash with 1 ml of RNase-free 1× PBS.

3.4 Hybridization of RNA-FISH Probe for Staining NEAT1_2

Before use, warm the NEAT1_2 Quasar 570 FISH probe stock solution to room temperature and briefly centrifuge.

To prepare the 125 nM hybridization buffer containing the NEAT1_2 FISH probe (see Sect. 2.4, step 1), add 1 μl of NEAT1_2 FISH probe stock solution to 50 μl of 10% hybridization buffer, mix gently, and centrifuge briefly (sufficient for one coverslip).

1. Aspirate the RNase-free 1× PBS from the coverslip-seeded U2OS NONO-GFP cells in the 12-well plate.
2. Add 500 μl of Wash buffer A and incubate at room temperature for 5 min.
3. In a humidified chamber (see Sect. 2.4, step 3), dispense 50 μl of the 125 nM of hybridization buffer containing the NEAT1_2 Quasar 570 FISH probe onto parafilm.

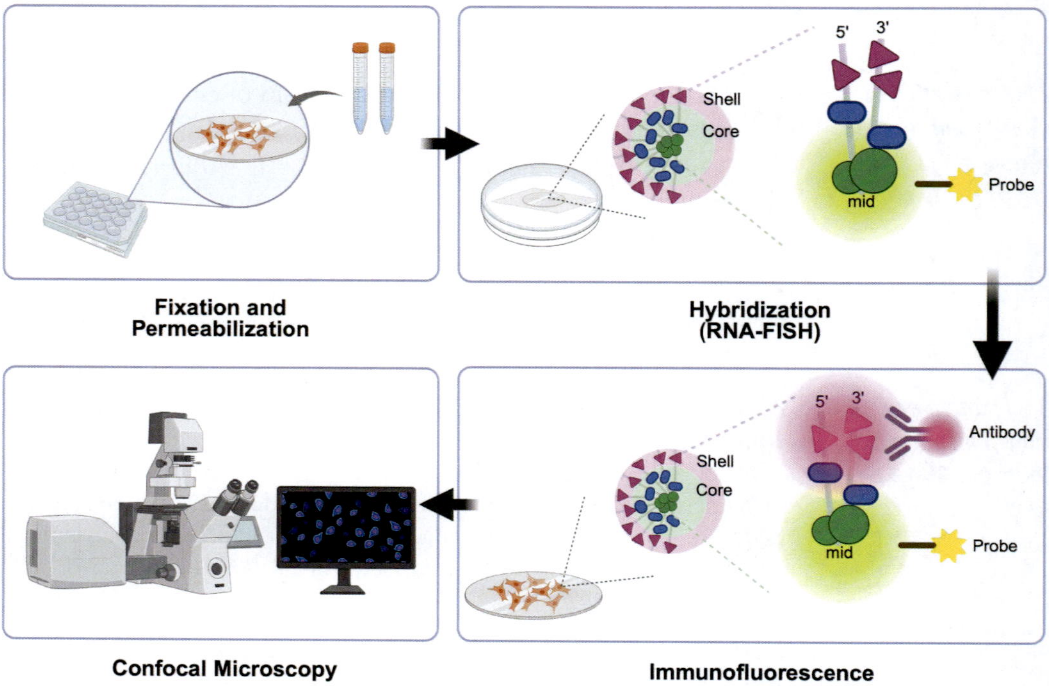

Fig. 2 Schematic procedure of sequential RNA-FISH and IF. Methodology for simultaneous visualization of paraspeckle core RNA and shell protein. Fixation and permeabilization preserve protein and RNA integrity and enable antibody access to the nucleus, while optimized blocking minimizes background for IF and prevents RNA-bound probe dissociation. High-resolution confocal microscopy is recommended to resolve the core–shell structure of paraspeckles

4. Using fine-tipped forceps, gently place the coverslip, cells facing down, onto the 50 μl drop of hybridization buffer containing the probe.
5. Cover the humidified chamber with its lid and seal with parafilm.
6. Incubate overnight in the dark at 37°C.
7. Using fine-tipped forceps, gently transfer the coverslip, cell facing up, into a fresh 12-well plate containing 500 μl of Wash buffer A.
8. Incubate in the dark at 37°C for 30 min.
9. Incubate with 500 μl of Wash buffer B at room temperature for 5 min.

3.5 Sequential IF for Staining TDP-43

To visualize NONO-GFP and the NEAT1_2 Quasar 570 FISH probe simultaneously, TDP-43 IF is performed using Alexa Fluor 647-conjugated secondary antibody. Depending on the staining combination, an alternative secondary antibody with a non-overlapping emission spectrum can be selected.

To prepare the TDP-43 primary antibody in blocking solution (see Sect. 2.5, step 2), dilute 1.2 μl of TDP-43 primary antibody in 360 μl of blocking solution (1:300 dilution). To prepare Alexa fluor 647 secondary antibody in blocking solution, dilute 1.6 μl of Alexa fluor 647 secondary antibody in 480 μl of blocking solution (1:300 dilution).

1. Wash with 1 ml of RNase-free 1× PBS for 5 min.
2. Block with 1 ml of blocking solution at room temperature for 30 min (*see* **Note 7**).
3. Remove the blocking solution and add diluted TDP-43 primary antibody in blocking solution.
4. Incubate overnight in the dark at 4°C.
5. Wash with RNase-free 1× PBS three times at room temperature for 5 min.
6. Incubate with the diluted Alexa Fluor 647 secondary antibody in blocking solution at room temperature for 1 h.
7. Wash with RNase-free 1× PBS three times at room temperature for 5 min.
8. Place a small drop of ProLong™ Gold Antifade Mountant onto a microscope slide, and mount a coverslip on the slide with cells facing down using fine-tipped forceps.
9. Allow to dry at room temperature in the dark for overnight.
10. Seal the perimeter of the coverslip with clear nail polish and store at 4°C under light-protected conditions.

3.6 Confocal Microscopy and Image Acquisition Parameters

Confocal imaging was performed using a Zeiss LSM980 microscope equipped with a Plan-Apochromat 63x/1.40 Oil DIC M27 objective lens.

Fluorescence images of the core–shell structure in paraspeckles are shown in Fig. 3.

1. Images were acquired at 2048 × 2048 pixels with 16-bit depth and a scan zoom of 2×.
2. To visualize NONO-GFP, NEAT1_2, and TDP-43, excitation lasers at 488, 553, and 653 nm were applied with a pinhole size of 1 Airy Unit (AU).
3. Each channel was scanned bidirectionally with a pixel dwell time of 2.05 μs.
4. Laser intensities were set to 0.3% (488 nm), 1.2% (553 nm), and 0.1%–0.2% (653 nm).
5. Detector gains were adjusted to 500 V (488 nm), 600 V (553 nm), and 520 V (653 nm).
6. Digital gains were set to 0.7 (488 nm), 0.7 (553 nm), and 0.5 (653 nm).

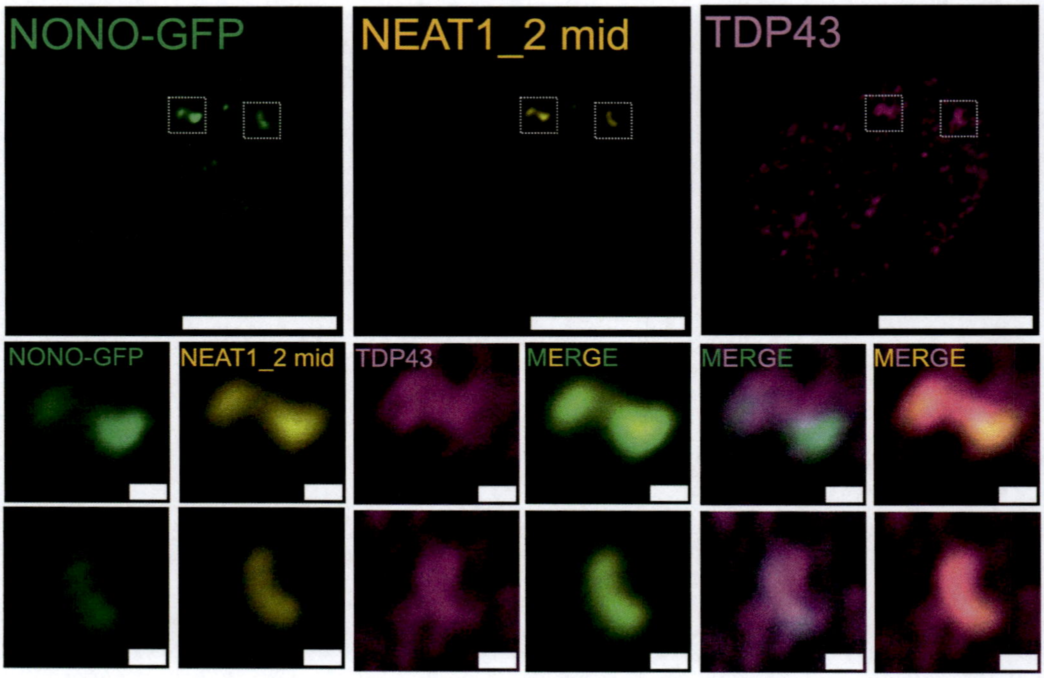

Fig. 3 Representative images showing the paraspeckle core–shell structure.U2OS cells stably expressing NONO-GFP were used to detect the paraspeckle core, and sequential RNA-FISH and IF were employed to visualize a core RNA component, the mid region of NEAT1_2, and a shell protein, TDP-43, respectively. GFP-tagged NONO and FISH probe-bound NEAT1_2 occupy the inner core within paraspeckles, and immunoreactive TDP-43 embraces the core. White boxes in the upper images indicate the magnified regions shown below. Scale bars: 10 μm (upper) and 0.5 μm (below)

4 Notes

1. pEGFP-N1-NONO construct was generated by in-house cloning. NONO (NM_001145408.2) was amplified from cDNA synthesized using RNA extracted from human embryonic kidney HEK293T cells. The amplified gene was cloned into the pEGFP-N1 vector through digestion with *Xho*I and *Bam*HI, followed by ligation with T4 DNA ligase.
2. To minimize RNase contamination, RNase-free solutions were prepared using 0.1% diethyl pyrocarbonate (DEPC)-treated distilled water (DW). DEPC-treated DW was generated by incubating DW with 0.1% DEPC (#D5758, Sigma-Aldrich) at room temperature for at least 4 h, followed by autoclaving at 121°C for 15 min.
3. To re-dissolve the dried FISH probe and create 12.5 μM FISH probe stock, blend the probe in TE buffer (10 mM Tris-HCl, 1 mM EDTA, pH 8.0) at the manufacturer's recommended

volume. Mix well by pipetting up and down, and then vortex and centrifuge briefly. The FISH probe stock should be stored at −20°C, protected from light, and subjected to minimal freeze-thaw cycles.

4. To obtain images of shell proteins surrounding the paraspeckle core structure, the use of a high-resolution confocal microscope is recommended.
5. Since the use of formaldehyde-containing methanol for fixation may affect the subsequent immunofluorescence following RNA-FISH, it is recommended to employ a fixation method without methanol.
6. As the type of permeabilization reagent, its concentration, and the duration of treatment may influence the immunofluorescence signal of nuclear proteins, it is recommended that identical conditions be employed to ensure reproducibility.
7. For preservation of RNA-FISH signal, a lower concentration and shorter duration of BSA blocking are recommended, as high-concentration solutions and prolonged blocking can interfere with the signals obtained in the preceding step.

Acknowledgments

This work is supported by grants from the National Research Foundation of Korea (NRF), funded by the Korean government (MSIT) (RS-2024-00408712) and the Ministry of Education through the Research Subsidies for Ph.D. Candidates (RS-2025-25426280).

References

1. Naganuma T, Nakagawa S, Tanigawa A, Sasaki YF, Goshima N, Hirose T (2012) Alternative 3′-end processing of long noncoding RNA initiates construction of nuclear paraspeckles. EMBO J 31:4020–4034
2. Yamazaki T, Souquere S, Chujo T, Kobelke S, Chong YS, Fox AH, Bond CS, Nakagawa S, Pierron G, Hirose T (2018) Functional domains of NEAT1 architectural lncRNA induce paraspeckle assembly through phase separation. Mol Cell 70:1038–1053 e1037
3. Sunwoo H, Dinger ME, Wilusz JE, Amaral PP, Mattick JS, Spector DL (2009) MEN epsilon/beta nuclear-retained non-coding RNAs are up-regulated upon muscle differentiation and are essential components of paraspeckles. Genome Res 19:347–359
4. Takakuwa H, Yamazaki T, Souquere S, Adachi S, Yoshino H, Fujiwara N, Yamamoto T, Natsume T, Nakagawa S, Pierron G, Hirose T (2023) Shell protein composition specified by the lncRNA NEAT1 domains dictates the formation of paraspeckles as distinct membraneless organelles. Nat Cell Biol 25: 1664–1675
5. Souquere S, Beauclair G, Harper F, Fox A, Pierron G (2010) Highly ordered spatial organization of the structural long noncoding NEAT1 RNAs within paraspeckle nuclear bodies. Mol Biol Cell 21:4020–4027
6. Hirose T, Virnicchi G, Tanigawa A, Naganuma T, Li R, Kimura H, Yokoi T, Nakagawa S, Benard M, Fox AH, Pierron G (2014) NEAT1 long noncoding RNA regulates transcription via protein sequestration within subnuclear bodies. Mol Biol Cell 25:169–183
7. Yamazaki T, Yamamoto T, Yoshino H, Souquere S, Nakagawa S, Pierron G, Hirose T

(2021) Paraspeckles are constructed as block copolymer micelles. EMBO J 40:e107270
8. West JA, Mito M, Kurosaka S, Takumi T, Tanegashima C, Chujo T, Yanaka K, Kingston RE, Hirose T, Bond C et al. (2016) Structural, super-resolution microscopy analysis of paraspeckle nuclear body organization. J Cell Biol 214:817–830
9. Okuda EK, Kessler LF, Arnold B, Riegger RJ, Hernandez Canas MC, Zebrowska E, Bakisoglu C, Rudigier M, Krost C, Nagasse HY et al. (2025) Rapid depletion and super-resolution microscopy reveal dual roles of SRSF5 in coordinating nuclear speckle-paraspeckle crosstalk during cellular stress. Nucleic Acids Res 53:gkaf713
10. Snead WT, Skillicorn MK, Shrinivas K, Gladfelter AS (2025) Immiscible proteins compete for RNA binding to order condensate layers. Proc Natl Acad Sci USA 122:e2504778122
11. Jiang L, Shao C, Wu QJ, Chen G, Zhou J, Yang B, Li H, Gou LT, Zhang Y, Wang Y et al. (2017) NEAT1 scaffolds RNA-binding proteins and the microprocessor to globally enhance pri-miRNA processing. Nat Struct Mol Biol 24:816–824
12. Zhang Y, Luo M, Cui X, O'Connell D, Yang Y (2022) Long noncoding RNA NEAT1 promotes ferroptosis by modulating the miR-362-3p/MIOX axis as a ceRNA. Cell Death Differ 29:1850–1863
13. Chen LL, DeCerbo JN, Carmichael GG (2008) Alu element-mediated gene silencing. EMBO J 27:1694–1705
14. Torres M, Becquet D, Blanchard MP, Guillen S, Boyer B, Moreno M, Franc JL, Francois-Bellan AM (2016) Circadian RNA expression elicited by 3′-UTR IRAlu-paraspeckle associated elements. Elife 5:e14837
15. Jacq A, Becquet D, Boyer B, Guillen S, Bello-Goutierrez MM, Blanchard MP, Villard C, Belghazi M, Torres M, Franc JL, Francois-Bellan AM (2025) Nuclear retention of mRNAs through paraspeckle protein binding to a sequence determinant in 3′UTR. Int J Mol Sci 26:6488

Check for updates

Chapter 11

Detection of Long Non-Coding RNA Localization by RNA Fluorescence In Situ Hybridization (FISH)

Hye-Rim Shin and Hyun Jung Oh

Abstract

Long non-coding RNAs (lncRNAs) regulate a wide range of important biological processes through their RNA molecules, yet the majority remain poorly understood at the functional level. Because subcellular localization plays a critical role in lncRNA function, determining spatial distribution is essential for understanding their mechanisms of action. To this end, RNA fluorescence in situ hybridization (RNA FISH) enables direct visualization of RNA molecules with single-cell spatial resolution. Here, we present a robust RNA FISH protocol for detecting the lncRNA *Xist*, a chromatin-associated master regulator of X chromosome inactivation (XCI) that serves as a paradigm for lncRNA biology. Using this protocol, we observe the characteristic *Xist* clouds on the inactive X chromosomes in mouse embryonic fibroblasts (MEFs), demonstrating the effectiveness of this method for investigating lncRNA localization and chromatin-associated function.

Keywords Long non-coding RNA, RNA fluorescence in situ hybridization (RNA FISH), RNA localization, Xist RNA, X chromosome inactivation (XCI), Chromatin

1 Introduction

Long non-coding RNAs (lncRNAs), generally defined as non-coding RNAs longer than 200 nucleotides, play roles in diverse processes ranging from chromatin regulation to transcriptional and post-transcriptional control [1]. Despite their biological importance, most lncRNAs remain functionally uncharacterized. LncRNAs function through the RNA molecule itself, and their subcellular localization—whether in the cytoplasm, nucleus, other organelles, or membrane-less compartments—is directly linked to their biological function [2–5]. Defining where an lncRNA resides is therefore an essential step toward deciphering its mechanisms of action.

Multiple experimental approaches have been established to determine RNA subcellular localization, including biochemical cell fractionation and proximity labeling techniques [4, 5]. Importantly, RNA fluorescence in situ hybridization (RNA FISH) offers distinct

Je-Hyun Yoon (ed.), *RNA Labeling*, Methods in Molecular Biology, vol. 3047,
https://doi.org/10.1007/978-1-0716-5352-4_11,

advantages by enabling spatial visualization of RNA molecules within intact cellular architecture at single-cell resolution (Fig. 1) [6, 7]. In our previous work, RNA FISH for the lncRNA *Jpx* revealed its diffusible nuclear distribution with focal accumulation, indicating that *Jpx* RNA translocates from its nascent transcription site on the X chromosomes to autosomes [8]. This result was consistent with *Jpx*'s genome-wide occupancy demonstrated by capture hybridization analysis of RNA targets sequencing (CHART-seq) [8]. RNA FISH therefore provides critical spatial context that complements biochemical and genomic approaches, enabling comprehensive characterization of lncRNA regulatory functions.

In RNA FISH, fluorescently labeled probes hybridize to complementary RNA sequences in fixed and permeabilized cells. Probe design strategies vary, with two major approaches: (1) long probes generated from DNA clones, such as bacterial artificial chromosomes (BACs), P1 plasmids, and cDNA plasmids, through nick translation and (2) probe sets of short, directly labeled oligonucleotides. The oligonucleotide-based approaches offer several advantages, including easier probe preparation and higher signal-to-noise ratios [6, 9, 10]. Single-molecule RNA-FISH (smFISH) represents a key advancement in oligonucleotide-based methods [9, 10]. Early smFISH protocols employed a small number of relatively long oligonucleotides, each labeled with multiple fluorophores [9]. Later refinements used many shorter oligonucleotide probes, each singly labeled with one fluorophore, that collectively tile along the target RNA [10]. This design improves specificity because individual off-target probe binding produces minimal fluorescence while on-target binding of multiple probes generates detectable signals. As a result, smFISH can resolve information obscured in ensemble measurements, including the stoichiometry and dynamics of individual transcripts as well as cell-to-cell heterogeneity in RNA abundance and localization [9–12].

Despite recent technical innovations [13], successful RNA FISH requires careful optimization for specific target RNAs and cell types. In this chapter, we present a detailed RNA FISH method to detect *Xist*, the lncRNA that drives X chromosome inactivation (XCI) to achieve dosage compensation between females and males [14, 15], in mouse embryonic fibroblasts (MEFs). *Xist* RNA exemplifies both the technical challenges and the power of RNA FISH for detecting chromatin-bound lncRNAs. During XCI, *Xist* spreads along and coats the future inactive X chromosome, forming a distinct *Xist* cloud [16–18]. Visualizing this process is critical for understanding XCI dynamics, *Xist*-mediated chromatin silencing, *Xist* compartment formation, and lncRNA–chromosome interactions. Both DNA clone-derived probes and fluorescently labeled oligonucleotides have been successfully applied to *Xist* RNA detection [8, 16–21]. Here, we describe both probe preparation methods but focus on the oligonucleotide-based approach, which we demonstrate with

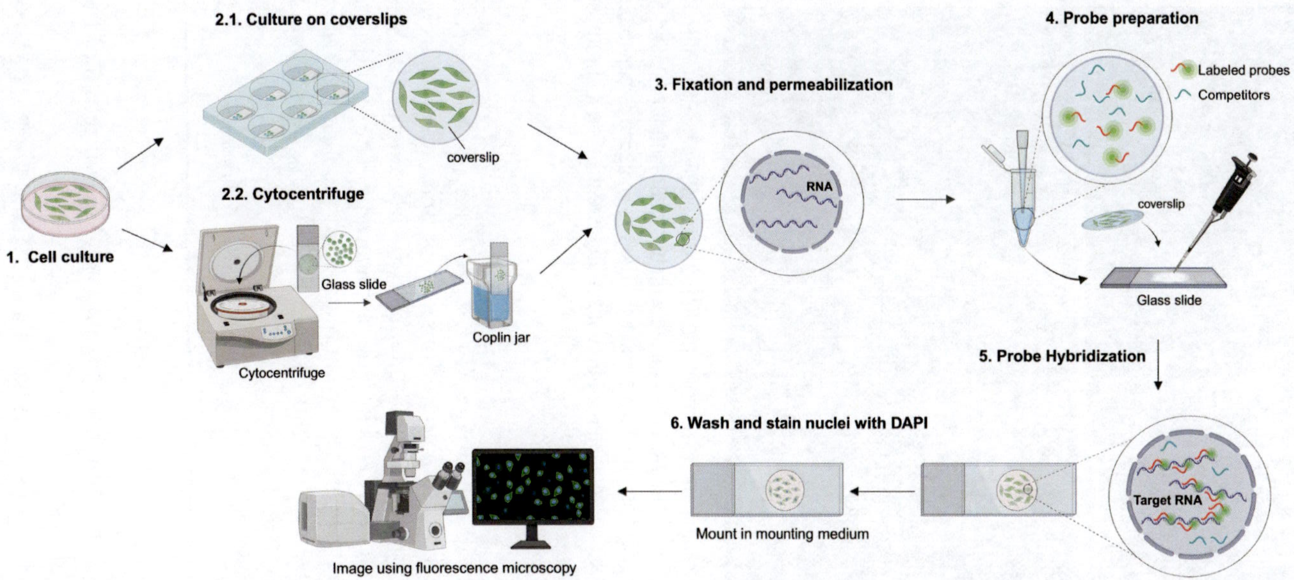

Fig. 1 Schematic representation of RNA FISH methodology. Cells are first prepared by either direct culture on coverslips or cytocentrifuge deposition onto glass slides. Cells are subsequently fixed and permeabilized to allow probe accessibility. Fluorescently labeled probes are prepared in hybridization buffer, with competitors such as Cot-1 DNA or yeast tRNA included to reduce nonspecific signals. During hybridization, probes specifically bind to complementary target RNA sequences. Following hybridization, samples are washed to remove unbound probes and mounted in antifade medium containing DAPI. RNA localization and abundance are then visualized by fluorescence microscopy, enabling in situ detection of transcripts such as long non-coding RNAs (lncRNAs). (Created in BioRender. SHIN, H. (2025) https://BioRender.com/mv7s9ll)

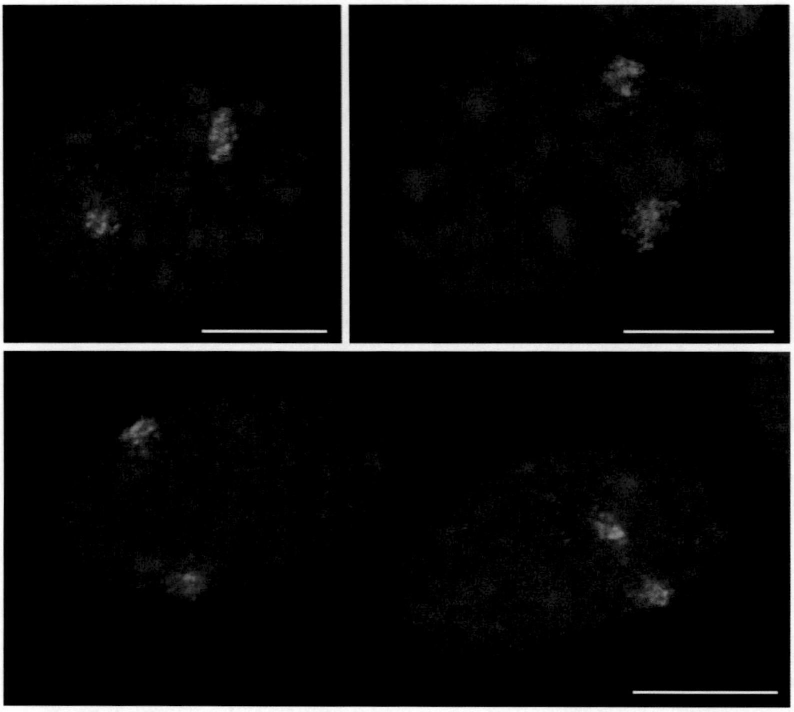

Fig. 2 Detection of *Xist* RNA foci in tetraploid mouse embryonic fibroblasts (MEFs) by RNA FISH. Representative *Xist* RNA FISH images of tetraploid MEFs. Cells were hybridized with two fluorescently labeled oligonucleotide probes (125 nM) against *Xist* RNA (green) and counterstained with DAPI to visualize nuclei (blue). Two distinct *Xist* RNA clouds were detected per nucleus, consistent with *Xist* coating two inactive X chromosomes in tetraploid cells. Scale bars, 10 μm

validated FISH images (Fig. 2). We provide optimized fixation and permeabilization conditions that preserve nuclear architecture and ensure probe accessibility, along with practical notes to support reproducibility and broader applications. The resulting fluorescence patterns reveal *Xist* accumulation on the inactive X chromosomes, highlighting the utility of RNA FISH for dissecting lncRNA function in chromatin biology.

2 Materials

2.1 Cell Culture

1. Mouse embryonic fibroblasts (MEFs): tetraploid transformed MEFs [22].
2. Dulbecco's modified Eagle Medium (DMEM; #10569044, Thermo Fisher Scientific) (*see* **Note 1**).
3. Fetal bovine serum (FBS; #F2442, Sigma-Aldrich).
4. 25 mM HEPES (#15630130, Thermo Fisher Scientific).

5. 1× MEM NEAA (#11140050, Thermo Fisher Scientific).
6. 1× Penicillin-Streptomycin (#15140163, Thermo Fisher Scientific).
7. 0.1 mM β-mercaptoethanol (#21985023, Thermo Fisher Scientific).
8. 100 mm culture plates.

2.2 Sample Preparation

Common Materials

1. 1× phosphate-buffered saline (PBS), pH 7.4.
2. 0.05% Trypsin-EDTA solution (#25300054, Thermo Fisher Scientific).
3. Cell counter (hemocytometer or automated cell counter).

2.2.1 Culture on Coverslips

1. 18 mm round glass coverslips (No. 1.5; #72222-01, Electron Microscopy Sciences) (*see* **Note 2**).
2. 12-well cell culture plates.
3. 0.2% (w/v) gelatin (#G1980, Sigma-Aldrich) solution; allow to dissolve completely after autoclaving.

2.2.2 Cytocentrifuge

1. Cytocentrifuge (e.g., Cytospin4, Epredia).
2. Cytoclips and cytofunnels with filter cards.
3. Positively charged glass slides (e.g., #22037246, Thermo Fisher Scientific) (*see* **Note 3**).
4. Coplin jar (#WH-900520, Wheaton) (*see* **Note 4**).

2.3 Fixation and Permeabilization

1. 10× PBS, pH 7.4.
2. Nuclease-free water (#10977015, Thermo Fisher Scientific).
3. Cytoskeletal (CSK) buffer; pH 6.8, filter-sterilize and store at 4°C.
   100 mM NaCl.
   300 mM sucrose.
   10 mM PIPES.
   3 mM $MgCl2{\cdot}6H2O$.
4. Triton X-100 (#T8787, Sigma-Aldrich).
5. 16% (w/v) paraformaldehyde (PFA) solution (#28908, Thermo Fisher Scientific) (*see* **Note 5**).
6. Ethanol.

2.4 Probe Preparation

Common Materials

1. Formamide, deionized (#F9037, Sigma-Aldrich) (*see* **Note 5**).
2. 20× SSC, pH 7.0 (#15557044, Thermo Fisher Scientific).
3. Dextran sulfate (50% solution; #3730, Millipore).
4. Ribonucleoside Vanadyl Complex (#S1402S, NEB).

5. Competitors (*see* **Note 6**)
   - Mouse Cot-1 DNA (#18330016, Thermo Fisher Scientific; supplied at 1 μg/μL).
   - Yeast tRNA (#AM7119, Thermo Fisher Scientific; supplied at 10 μg/μL) .
6. Ethanol.
7. Nuclease-free water (#10977015, Thermo Fisher Scientific).
8. Thermal cycler.

2.4.1 DNA Clone-Derived Probes Through Nick Translation

1. DNA clones (1 μg per reaction).
2. Fluorescently labeled dUTP (e.g., Fluorescein-12-dUTP; #1137324910, Sigma-Aldrich).
3. dATP, dGTP, dCTP, and dTTP (#R0181, Thermo Fisher Scientific).
4. Nick translation enzyme mix
   - 1000 U/mL DNA polymerase I (#M2051, Promega).
   - 3 U/mL RQ1 RNase-Free DNase (#M6101, Promega).
   - 1× DNA polymerase I buffer (#M2051, Promega).
   - 0.5 mg/mL BSA (#AM2616, Thermo Fisher Scientific).
   - 50% glycerol.
   - *Prepare 500 μL in nuclease-free water, aliquot, and store at −20°C.*
5. 0.5 M EDTA, pH 8.0.
6. Linear acrylamide (#AM9520, Thermo Scientific).
7. 3 M Sodium acetate, pH 5.5.
8. Hybridization buffer (*see* **Note 7**)
   - 50% Formamide, 2× SSC.
   - 10% Dextran sulfate.
   - 2 mM Ribonucleoside Vanadyl Complex .
   - *Use freshly prepared or store −20°C.*

2.4.2 Fluorescently Labeled Oligonucleotide Probes

1. Fluorescently labeled oligonucleotide probes (example sequences for *Xist* [20]):
   - ATGACTCTGGAAGTCAGTATGGAG-6-FAM.
   - ATGGGCACTGCATTTTAGCAATA-6-FAM.
2. Hybridization buffer (*see* **Note 7**)
   - 30% Formamide, 2× SSC.
   - 10% Dextran sulfate.

- 2 mM Ribonucleoside Vanadyl Complex .
- *Use freshly prepared or store −20°C.*

2.5 Probe Hybridization

1. 80%, 90%, 100% Ethanol.
2. Glass slides for coverslip culture-based preparation (#22037246, Thermo Fisher Scientific).
3. Glass coverslips for cytocentrifuge-based preparation (#72222-01, Electron Microscopy Sciences).
4. Rubber cement (for sealing coverslips).
5. Humidified chamber.
6. 37°C incubator.
7. Thermal Cycler.

2.6 Wash and Mounting

1. Wash buffer I for DNA clone-derived probes: 50% formamide in 2× SSC.
2. Wash buffer I for fluorescently labeled oligonucleotide probes: 30% formamide in 2× SSC.
3. Wash buffer II: 2× SSC.
4. Mounting medium containing DAPI (#P36966, Thermo Fisher Scientific).
5. Confocal microscope (e.g., LSM 900, ZEISS) or widefield fluorescence microscope.

3 Methods

3.1 Cell Culture

Optimal growth conditions are critical for maintaining cellular integrity and RNA quality.

1. Culture cell lines (e.g., MEFs) in supplemented DMEM (Sect. 2.1; *see* **Note 1**).
2. Culture cells until reaching approximately 80%–90% confluence.

3.2 Sample Preparation

Direct culture on coverslips preserves cell morphology and nuclear structure. Cytocentrifugation offers an alternative method to uniformly deposit cells onto slides when direct culture is impractical.

3.2.1 Culture on Coverslips

1. Sterilize coverslips (e.g., 18 mm round) by autoclaving.
2. Place sterile coverslips in individual wells of a 12-well culture plate and coat with 0.2% (w/v) gelatin solution for 30 min at room temperature.
3. Remove excess gelatin solution.

4. Wash cells with 1× PBS, trypsinize with 0.05% Trypsin-EDTA solution, and count cells.
5. Adjust cell concentration to $1.0–1.2 \times 10^5$ cells/mL and seed 1 mL of the cell suspension into each gelatin-coated coverslip.
6. Culture cells until reaching approximately 80% confluence.

3.2.2 Cytocentrifuge

1. Wash cells with 1× PBS, trypsinize with 0.05% Trypsin-EDTA solution, and count cells.
2. Adjust cell concentration and resuspend in 1× PBS to achieve a final concentration of 5×10^5 cells/mL.
3. Transfer 200 μL of cell suspension into each cytofunnel containing glass slides (yielding 1×10^5 cells per funnel) (*see* **Note 8**).
4. Perform cytocentrifugation at 113 × g (1000 rpm) for 5 min at room temperature (*see* **Note 8**).
5. Carefully remove the slides from the cytocentrifuge and allow them to air-dry briefly.

3.3 Fixation and Permeabilization

Fixation and permeabilization preserve RNA localization and cellular structure while allowing probe penetration.

1. Immerse coverslips or slides attached with cells in ice-cold 1× PBS for 2 min.
2. Remove PBS and permeabilize cells with ice-cold cytoskeletal (CSK) buffer supplemented with 0.5% (v/v) Triton X-100 for optimized time (e.g., ~5 min for MEFs; *see* **Note 9**).
3. Fix cells in freshly prepared 4% (w/v) paraformaldehyde (PFA) in 1× PBS for 10 min at room temperature, prepared by diluting a 16% (w/v) PFA stock solution with 10× PBS and nuclease-free water to yield final concentrations of 4% PFA and 1× PBS.
4. Wash samples with 1× PBS for 1 min, then immerse samples in 70% (v/v) ethanol for 2 min. Samples can be stored in 70% (v/v) ethanol at 4°C or −20°C until further processing, or proceed immediately to Sect. 3.5.

3.4 Probe Preparation

High-quality probe preparation ensures optimal hybridization specificity and sensitivity. Probe preparation can be performed in advance and independently of sample processing (Sects. 3.1–3.3). Prepared probes can be stored at −20°C until use.

3.4.1 DNA Clone-Derived Probes Through Nick Translation

1. Assemble nick translation reaction on ice in a total volume of 50 μL:
   - 1 μg DNA clone (e.g., pSx9 plasmid for *Xist* RNA FISH [8, 19]).
   - 3 μL 0.2 mM fluorescence-labeled dUTP.
   - 2 μL 0.2 mM dTTP.

- 5 μL 0.2 mM dGTP, 0.2 mM dCTP, 0.2 mM dATP.
- 5 μL 10X DNA polymerase I buffer.
- 5 μL nick translation enzyme mix.
- Add nuclease-free water to adjust the final volume to 50 μL.

2. Perform nick translation in a thermal cycler at 15°C for 90 min, followed by enzyme inactivation at 75°C for 15 min. Hold at 4°C (*see* **Note 10**).
3. Briefly centrifuge, transfer reaction to a 1.5 mL RNase-free microcentrifuge tube, and add sequentially:
   - 1 μL of 0.5 M EDTA.
   - 1 μL of linear acrylamide.
   - 5 μL of 3 M sodium acetate (pH 5.5).
   - 125 μL of ice-cold 100 % ethanol.
4. Precipitate at −80°C for 2 h, then centrifuge at maximum speed at 4°C for 30 min.
5. Wash pellet with 1 mL of 70% (v/v) ethanol and centrifuge at maximum speed at 4°C for 5 min.
6. Air-dry pellet and dissolve in 10 μL nuclease-free water, yielding 100 ng/μL.
7. Verify probe quality by running 1 μL on a 2% agarose gel (*see* **Note 11**).
8. Prepare probe working solution by diluting labeled probes (e.g., 2 ng/μL for *Xist*) in hybridization buffer containing mouse Cot-1 DNA (e.g., 40 ng/μL) as competitor for probes with repetitive sequences (*see* **Note 12**).
   - *Alternative approach*: to streamline the workflow, add the desired amount of Cot-1 DNA to the precipitation mix at Step 3 to co-precipitate with probes, then dissolve the pellet directly in hybridization buffer at working concentrations (Step 6).

3.4.2 Fluorescently Labeled Oligonucleotide Probes

1. Prepare 100 μM stock solutions (100 pmol/μL) of each fluorescently labeled *Xist* oligonucleotide probe in nuclease-free water.
2. Combine equal volumes of the two probes to yield a 100 μM mixture (~788 ng/μL total).
3. Prepare probe working solution by diluting labeled probes in hybridization buffer containing 100 ng/μL mouse Cot-1 DNA or 1 μg/μL yeast tRNA as competitor. Concentrations as low as 125 nM (~1 ng/μL) provide successful detection, with higher concentrations up to 1.25 μM (~10 ng/μL) also tested (*see* **Note 13**).

3.5 Probe Hybridization

Hybridization allows specific binding of fluorescent probes to complementary RNA sequences within cells.

1. Incubate probe working solution in thermal cycler at 80°C for 10 min and subsequently at 37°C for 5–10 min.
2. During probe pre-incubation, complete the dehydration series (from 70% ethanol; Sect. 3.3) by sequential immersion in 80%, 90%, and 100% ethanol for 2 min each.
3. Air-dry samples at RT for <5 min.
4. Apply 6–8 μL of the probe-competitor mixture in hybridization buffer to each sample area.
5. Place coverslips cell-side down on slides for coverslip-cultured samples, or place a coverslip over each cytospun slide for cytocentrifuged samples.
6. Seal edges with rubber cement and incubate in a dark humidified chamber at 37°C overnight (*see* **Note 14**).

3.6 Wash and Mounting

Post-hybridization washes eliminate unbound probes to minimize background signal. Mounting medium with DAPI enables nuclear visualization and sample preservation.

1. Carefully remove rubber cement to unseal edges, then gently separate slide and coverslip.
2. Wash samples three times for 5 min each in wash buffer I at 37°C in the dark (*see* **Note 15**).
3. Wash samples three times for 5 min each in wash buffer II at 37°C in the dark (*see* **Note 15**).
4. Air-dry samples for <2 min, then apply 6–8 μL of mounting medium containing DAPI between slide and coverslip.
5. Image slides using fluorescence or confocal microscopy.

4 Notes

1. Adjust media and supplement composition based on specific requirements for each cell line being used.
2. Choose a coverslip diameter according to the culture well size.
3. Positively charged slides enhance cellular attachment during cytocentrifugation and subsequent processing.
4. Coplin jars enable convenient handling for slides with cytospun cells.
5. Handle paraformaldehyde solution and formamide in a fume hood.

6. Nonspecific competitors improve signal specificity. Cot-1 DNA is particularly recommended for probes containing repetitive sequences. Concentrate stock solutions if needed: precipitate competitors using ethanol and dissolve the resulting pellets directly in hybridization buffer.
7. The formamide concentration is probe dependent (30% formamide for *Xist* oligonucleotide probes; 50% formamide for DNA clone-derived probes). For other targets, empirical optimization is required, as higher concentrations reduce DNA–RNA hybrid stability and melting temperature (Tm).
8. Cell density and centrifugation parameters may require adjustment for different cell types to ensure uniform monolayer distribution and minimize cell loss.
9. Permeabilization duration varies by cell type and experimental condition. For example, MEFs require ~5 min. Fixation can be performed before permeabilization depending on cell type and target.
10. Prolonged nick translation causes excessive probe fragmentation. Reaction duration may require adjustment for different BAC or plasmid templates.
11. Expected probe size by electrophoresis: smear pattern ranging from 50 to 500 bp, with maximum intensity at ~200 bp.
12. The probe-to-Cot-1 ratio may require optimization for different targets. Test different concentrations of target probes at 2, 5, or 10 ng/μL with Cot-1 DNA at various excess ratios.
13. Probe concentration may require optimization for different targets.
14. Hybridization temperature and duration are probe dependent. The conditions shown are optimized for *Xist* probes; other targets may require adjustments (e.g., oligonucleotide probes: room temperature to 37°C; DNA clone-derived probes: 37–42°C; duration: several hours to overnight).
15. Adjust the wash temperature based on the hybridization temperature.

Acknowledgments

This work was supported by the Yonsei University Research Fund (No. 2024–22–0520 to H.J.O.); by National Research Foundation of Korea grants, funded by the Ministry of Science and ICT (Nos. RS–2024–00342549 and RS–2024–00403165 to H.J.O.); by Global-Learning & Academic research institution for Master's·PhD students and Postdocs (LAMP) Program of the National Research

Foundation of Korea grant, funded by the Ministry of Education (No. RS–2024–00442483 to H.-R.S. and H.J.O.); and by a POSCO Science Fellowship by the POSCO TJ Park Foundation (to H.J.O.)

References

1. Mattick JS, Amaral PP, Carninci P, Carpenter S, Chang HY, Chen LL et al. (2023) Long non-coding RNAs: definitions, functions, challenges and recommendations. Nat Rev Mol Cell Biol 24(6):430–447. https://doi.org/10.1038/s41580-022-00566-8
2. Bridges MC, Daulagala AC, Kourtidis A (2021) LNCcation: lncRNA localization and function. J Cell Biol 220(2):e202009045. https://doi.org/10.1083/jcb.202009045
3. Chen LL (2016) Linking long noncoding RNA localization and function. Trends Biochem Sci 41(9):761–772. https://doi.org/10.1016/j.tibs.2016.07.003
4. Carlevaro-Fita J, Johnson R (2019) Global positioning system: understanding long noncoding RNAs through subcellular localization. Mol Cell 73(5):869–883. https://doi.org/10.1016/j.molcel.2019.02.008
5. Biayna J, Dumbovic G (2025) Decoding subcellular RNA localization one molecule at a time. Genome Biol 26(1):45. https://doi.org/10.1186/s13059-025-03507-8
6. Levsky JM, Singer RH (2003) Fluorescence in situ hybridization: past, present and future. J Cell Sci 116(Pt 14):2833–2838. https://doi.org/10.1242/jcs.00633
7. Singer RH, Ward DC (1982) Actin gene expression visualized in chicken muscle tissue culture by using in situ hybridization with a biotinated nucleotide analog. Proc Natl Acad Sci U S A 79(23):7331–7335. https://doi.org/10.1073/pnas.79.23.7331
8. Oh HJ, Aguilar R, Kesner B, Lee HG, Kriz AJ, Chu HP et al. (2021) Jpx RNA regulates CTCF anchor site selection and formation of chromosome loops. Cell 184(25):6157–6173.e24. https://doi.org/10.1016/j.cell.2021.11.012
9. Femino AM, Fay FS, Fogarty K, Singer RH (1998) Visualization of single RNA transcripts in situ. Science 280(5363):585–590. https://doi.org/10.1126/science.280.5363.585
10. Raj A, van den Bogaard P, Rifkin SA, van Oudenaarden A, Tyagi S (2008) Imaging individual mRNA molecules using multiple singly labeled probes. Nat Methods 5(10):877–879. https://doi.org/10.1038/nmeth.1253
11. Raj A, van Oudenaarden A (2009) Single-molecule approaches to stochastic gene expression. Annu Rev Biophys 38:255–270. https://doi.org/10.1146/annurev.biophys.37.032807.125928
12. Kwon S (2013) Single-molecule fluorescence in situ hybridization: quantitative imaging of single RNA molecules. BMB Rep 46(2):65–72. https://doi.org/10.5483/bmbrep.2013.46.2.016
13. Le P, Ahmed N, Yeo GW (2022) Illuminating RNA biology through imaging. Nat Cell Biol 24(6):815–824. https://doi.org/10.1038/s41556-022-00933-9
14. Lee JT (2011) Gracefully ageing at 50, X-chromosome inactivation becomes a paradigm for RNA and chromatin control. Nat Rev Mol Cell Biol 12(12):815–826. https://doi.org/10.1038/nrm3231
15. Dossin F, Heard E (2022) The molecular and nuclear dynamics of X-chromosome inactivation. Cold Spring Harb Perspect Biol 14(4):a040196. https://doi.org/10.1101/cshperspect.a040196
16. Clemson CM, McNeil JA, Willard HF, Lawrence JB (1996) XIST RNA paints the inactive X chromosome at interphase: evidence for a novel RNA involved in nuclear/chromosome structure. J Cell Biol 132(3):259–275. https://doi.org/10.1083/jcb.132.3.259
17. Sunwoo H, Wu JY, Lee JT (2015) The Xist RNA-PRC2 complex at 20-nm resolution reveals a low Xist stoichiometry and suggests a hit-and-run mechanism in mouse cells. Proc Natl Acad Sci U S A 112(31):E4216–E4225. https://doi.org/10.1073/pnas.1503690112
18. Markaki Y, Gan Chong J, Wang Y, Jacobson EC, Luong C, Tan SYX et al. (2021) Xist nucleates local protein gradients to propagate silencing across the X chromosome. Cell 184(25):6174–6192.e32. https://doi.org/10.1016/j.cell.2021.10.022
19. Cohen DE, Davidow LS, Erwin JA, Xu N, Warshawsky D, Lee JT (2007) The DXPas34 repeat regulates random and imprinted X inactivation. Dev Cell 12(1):57–71. https://doi.org/10.1016/j.devcel.2006.11.014
20. Del Rosario BC, Del Rosario AM, Anselmo A, Wang PI, Sadreyev RI, Lee JT (2017) Genetic intersection of Tsix and Hedgehog signaling during the initiation of X-chromosome inactivation. Dev Cell 43(3):359–371.e6. https://doi.org/10.1016/j.devcel.2017.09.027

21. Colognori D, Sunwoo H, Kriz AJ, Wang CY, Lee JT (2019) Xist deletional analysis reveals an interdependency between Xist RNA and polycomb complexes for spreading along the inactive X. Mol Cell 74(1):101–117.e10. https://doi.org/10.1016/j.molcel.2019.01.015

22. Yildirim E, Sadreyev RI, Pinter SF, Lee JT (2011) X-chromosome hyperactivation in mammals via nonlinear relationships between chromatin states and transcription. Nat Struct Mol Biol 19(1):56–61. https://doi.org/10.1038/nsmb.2195

Check for updates

Chapter 12

Labeling of Long Noncoding RNAs to Study Virus-Induced Nuclear Aggregates

Eui Tae Kim

Abstract

Nuclear RNA-protein condensates, including nuclear speckles and paraspeckles, are highly dynamic subnuclear compartments that influence RNA processing and transcriptional control. During viral infection, particularly by herpes simplex virus 1 (HSV-1), these structures undergo marked reorganization. The long noncoding RNA NEAT1 is a key determinant of paraspeckle assembly and activity under such conditions. This chapter reviews current insights into how viruses manipulate or disrupt nuclear condensates and presents a practical protocol for visualizing NEAT1-positive paraspeckles in HSV-1-infected HeLa cells by RNA fluorescence in situ hybridization (RNA FISH). The protocol provides stepwise guidance for cell preparation, viral infection, probe labeling, and fluorescence microscopy, supplemented with troubleshooting notes to support reproducibility. These approaches allow sensitive detection of virus-driven nuclear RNA assemblies and enable detailed investigation of host-virus interactions within the nucleus.

Keywords Herpes simplex virus, Nuclear speckles, Paraspeckles, Long noncoding RNA, RNA FISH, NEAT1, smFISH

1 Introduction

In mammalian cell nuclei, a variety of membrane-less nuclear bodies (NBs) exist, including nuclear speckles, paraspeckles, nucleoli, Cajal bodies, and PML bodies. These structures concentrate proteins and RNAs to compartmentalize nuclear functions such as transcriptional regulation, RNA processing, and ribosome assembly. Nuclear speckles (also known as SC35 domains or interchromatin granule clusters) are enriched in pre-mRNA splicing factors and poly(A)+ RNA, serving as hubs for splicing and mRNA export [1, 2]. A key architectural component is the lncRNA MALAT1 (NEAT2), which localizes to speckles and contributes to their structural integrity and transcriptional regulation through interactions with chromatin modifiers such as EZH2 [3, 4]. Paraspeckles, in contrast, are built on the long NEAT1_2 isoform (~23 kb), which scaffolds proteins including

Je-Hyun Yoon (ed.), *RNA Labeling*, Methods in Molecular Biology, vol. 3047,
https://doi.org/10.1007/978-1-0716-5352-4_12,

PSPC1, NONO (p54nrb), and SFPQ (PSF) [5–8]. Typically, 5–20 paraspeckles are observed per nucleus, appearing as punctate foci in the interchromatin space [9]. They retain hyper-adenosine-inosine (A-to-I) edited RNAs and sequester transcriptional regulators to modulate gene expression [10]. Under stress conditions such as proteasome inhibition or viral infection, NEAT1 expression increases, leading to enhanced paraspeckle assembly [11, 12]. Viral infections can substantially alter the composition and function of nuclear bodies. Herpes simplex virus 1 (HSV-1), a nuclear DNA virus, induces NEAT1 expression and promotes the aberrant colocalization of the nuclear speckle protein SRSF2 with paraspeckle proteins (PSPC1 and NONO) within nuclear foci. These complexes localize to viral promoters and correlate with histone modifications mediated by EZH2 and CBP/p300 that favor transcription. Silencing of SRSF2 or paraspeckle components inhibits HSV-1 gene expression, underscoring their proviral role [2]. In parallel, HSV-1 disperses SRSF2/SC35-positive nuclear speckles, contributing to global inhibition of host splicing and export [13]. NEAT1 upregulation, driven by STAT3, further enhances paraspeckle formation and recruits STAT3 to viral promoters, supporting lytic gene transcription [14].

Other viruses engage paraspeckles in different ways. Influenza A virus (IAV) infection induces NEAT1 and paraspeckle assembly, sequestering SFPQ away from cytokine promoters such as IL-8, thereby enhancing immune gene expression [15]. SFPQ also interacts with influenza polymerase to facilitate viral mRNA polyadenylation, illustrating the dual impact of NEAT1 sequestration [16]. Hantaan virus similarly induces NEAT1 via RIG-I/IRF7, promoting paraspeckle formation that relocates SFPQ away from interferon-stimulated gene promoters (e.g., RIG-I, DDX60) to enhance their transcription and restrict viral spread [17].

In chronic infection, paraspeckles take on additional roles. In HIV-1, NEAT1 sequesters unspliced viral RNAs in paraspeckles, limiting cytoplasmic export and contributing to latency [18]. In contrast, MALAT1 displaces PRC2 from the HIV LTR promoter to reactivate proviral transcription [19]. Thus, paraspeckle-associated NEAT1 promotes latency while speckle-associated MALAT1 promotes reactivation, illustrating opposing functions of nuclear lncRNAs in HIV persistence.

Certain DNA viruses further reshape paraspeckles morphologically. Kaposi's sarcoma-associated herpesvirus (KSHV) forms enlarged paraspeckle-like condensates (v-mPS) during lytic replication, composed of NEAT1 and core paraspeckle proteins (SFPQ, PSPC1, and NONO) but lacking FUS [20]. These atypical structures localize near replication centers, facilitating viral RNA processing while also contributing to host genomic instability through sequestration of RNA-binding proteins. These virus-modified condensates suggest a potential role in oncogenic transformation, although further validation is needed.

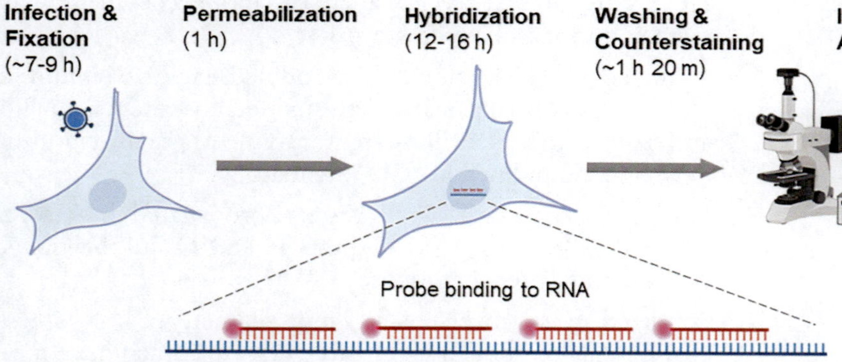

Fig. 1 Workflow of RNA FISH for detecting lncRNA in virus-infected cells. Schematic overview of the major steps in an RNA FISH experiment. Cells are infected or treated, fixed, and permeabilized, followed by hybridization of fluorescently end-labeled oligonucleotide probes to the target RNAs. After washing and counterstaining, RNA localization can be visualized and analyzed by fluorescence microscopy. The times indicated in parentheses are provided as a reference based on HSV-1 infection

Taken together, these findings demonstrate that nuclear speckles and paraspeckles act as either antiviral or proviral platforms depending on the pathogen. To investigate these interactions, sensitive methods such as RNA FISH are essential. RNA fluorescence in situ hybridization (RNA FISH), particularly single-molecule FISH (smFISH), enables sensitive detection of NEAT1 as paraspeckle foci. Here, we describe a protocol to label and image nuclear RNA aggregates in HSV-1-infected HeLa cells, focusing on NEAT1-containing paraspeckles. The protocol includes detailed steps for cell culture, HSV-1 infection, sample fixation, hybridization with a fluorescent NEAT1 probe set, and fluorescence imaging (Fig. 1). Notes are provided to discuss technical tips, such as maintaining RNase-free conditions and selecting appropriate probe target regions (NEAT1 isoforms). This approach enables researchers to directly observe virus-induced nuclear RNA granules and assess how viral infections perturb nuclear compartmentalization at the single-cell level.

2 Materials

All cell culture and HSV-1 infection procedures must be performed in a certified biosafety cabinet with appropriate containment (BSL-2). Infectious viruses should be handled according to institutional biosafety guidelines. All solutions and reagents contacting RNA should be prepared RNase-free (e.g., DEPC-treated water) to preserve RNA integrity prior to fixation.

2.1 Cell Culture and Virus Infection

1. HeLa cells (human cervical adenocarcinoma cell line; low passage; confirmed mycoplasma-free).
2. Complete DMEM: Dulbecco's Modified Eagle's Medium supplemented with 10% heat-inactivated fetal bovine serum (FBS) and 100 U/mL penicillin-streptomycin. Maintain cultures at 37°C in a humidified 5% CO_2 incubator.
3. HSV-1 stock (strain KOS or 17syn+, propagated in Vero cells; ~1×10^8 PFU/mL). Store aliquots at −80°C and thaw rapidly on ice immediately before use.
4. Sterilized glass coverslips (12 mm diameter). Poly-D-lysine coating may be applied if enhanced cell attachment is required.
5. Phosphate-buffered saline (PBS), pH 7.4 (RNase-free).

2.2 Fixation and Hybridization Reagents

1. Paraformaldehyde, 4% (w/v) in PBS. Prepare fresh on the day of use. Handle in a fume hood with gloves.
2. Ethanol, 70% (v/v) in nuclease-free water. Prechill at −20°C before use.
3. NEAT1 RNA FISH probe set (e.g., Stellaris® probes, Quasar® 570-labeled, targeting NEAT1_2). Stock: 250 μM in TE buffer. Store at −20°C, protected from light.
4. Hybridization buffer: 10% formamide, 2× SSC, 10% dextran sulfate (w/v), 1 mg/mL *E. coli* tRNA, 0.02% RNase-free BSA. Store at 4°C.
5. Wash Buffer A: 2× SSC with 10% formamide. Prewarm to 37°C before use.
6. Wash Buffer B: 2× SSC. Optional: supplement with 0.1% Triton X-100.
7. DAPI nuclear stain. Stock: 1 mg/mL in water. Working concentration: 1 μg/mL. Protect from light.
8. Anti-fade mounting medium (e.g., ProLong™ Gold or Fluoromount-G). Store at −20 °C and warm to room temperature before use.

2.3 Microscope and Imaging Supplies

1. Fluorescence microscope (widefield or confocal) equipped with a 60× or 100× oil-immersion objective and filter sets or lasers compatible with DAPI and Quasar 570 (TRITC/Cy3 range).
2. Immersion oil for oil immersion objectives, refractive index matched to the lens.
3. Microscope slides and coverslip tools. Clean glass slides, fine forceps, and nail polish or sealant to secure coverslips if required.

4. Image analysis software (optional). Fiji/ImageJ or equivalent for RNA foci counting and fluorescence intensity (*see* **Note 8**).
5. Reagent setup. Thaw HSV-1 stocks on ice before infection. Warm hybridization buffer and Wash Buffer A (formamide-containing) to 37°C. Protect NEAT1 FISH probes from light, thaw aliquots on ice just before use, and use RNase-free buffers (PBS, SSC).

3 Methods

The overall workflow involves seeding HeLa cells on coverslips, infecting them with HSV-1 for a defined period, fixing and permeabilizing, hybridizing with fluorescent NEAT1 probes, washing to remove unbound probes, and finally staining nuclei and mounting for imaging (Fig. 1). All steps involving samples post-fixation should be done in the dark or low-light to prevent photobleaching of fluorophores.

3.1 Cell Seeding and Viral Infection

1. 1 day prior to infection, seed ~5×10^4 HeLa cells onto sterile 12 mm coverslips in a 24-well plate with 0.5 mL complete DMEM. Incubate overnight to ~70% confluence (*see* **Note 1**).
2. Thaw one aliquot of HSV-1 stock on ice. Dilute in serum-free DMEM to MOI of ~5 (e.g., 5 μL of 1×10^8 PFU/mL stock in 0.5 mL medium, ~10^5 cells per coverslip). Mix gently.
3. Remove medium from cells, wash once with 1 mL prewarmed PBS, and add 0.5 mL virus inoculum per coverslip. For mock infection, add 0.5 mL serum-free medium only.
4. Incubate at 37°C for 1 h with gentle rocking every 15 min to facilitate uniform infection.
5. Aspirate inoculum, wash once with 1 mL warm PBS, then add 1 mL complete DMEM. Incubate 6–8 h at 37°C for NEAT1/paraspeckle analysis (*see* **Note 2**).
6. For longer infection (12–16 h), expect stronger viral gene expression but also host shutoff.
7. Optionally check cells under light microscopy during infection. HSV-1-infected HeLa cells show rounding and mild chromatin condensation by ~8 h. Confirm cells remain attached and not excessively lysed prior to fixation.

3.2 Fixation and Permeabilization

1. At ~8 h post-infection (or chosen timepoint), prepare samples for fixation in a fume hood to preserve RNA integrity and nuclear structure.

2. Remove culture medium, rinse each coverslip with 1 mL PBS, and add 0.5 mL freshly prepared 4% paraformaldehyde in PBS. Incubate 10 min at room temperature.
3. Remove the fixative and rinse twice with PBS. Optionally, quench with 0.1 M glycine in PBS for 5 min to reduce autofluorescence.
4. Add 1 mL cold (−20°C) 70% ethanol per well and incubate ≥ 1 h at 4°C. Coverslips can be stored in ethanol at 4°C for 1–14 days or at −20°C for long-term storage. At this stage, samples can be handled outside the biosafety cabinet but must remain RNase-free.
5. Before hybridization, replace ethanol with 0.5 mL Wash Buffer A (2× SSC, 10% formamide) prewarmed to 37°C, and incubate 5 min at room temperature to equilibrate for probe binding.

3.3 RNA FISH Hybridization (NEAT1 smFISH)

1. Thaw the NEAT1 Stellaris probe stock on ice and protect from light. Prepare 100 μL hybridization mix per coverslip by combining Hybridization Buffer with probe at a final concentration of 125 nM (e.g., 10 μL of a 12.5 μM working solution plus 90 μL buffer). Mix gently without vortexing (*see* **Note 3**).
2. Remove Wash Buffer A from the coverslips and add 100 μL hybridization mix directly, avoiding bubbles. Optionally, use Parafilm to minimize evaporation.
3. Place coverslips in a humidified chamber at 37°C and incubate overnight (~16 h). Maintain humidity with water or damp paper towels. Protect from light to prevent bleaching.
4. Following hybridization, wash coverslips in 500 μL Wash Buffer A (prewarmed to 37°C) for 30 min with gentle shaking to remove unbound probe while preserving RNA-probe hybrids.
5. Wash again in fresh Wash Buffer A for 15 min at 37°C, then transfer to Wash Buffer B (2× SSC without formamide) for 5 min at room temperature.
6. Add DAPI to Wash Buffer B (final 5–50 ng/mL) and incubate 5 min to stain nuclei. Rinse briefly in fresh 2× SSC to remove excess stain.
7. Mount coverslips on glass slides with 10–20 μL anti-fade medium. Invert coverslips (cell side down), avoid bubbles, and seal edges with nail polish or sealant. Cure in the dark for ≥ 24 h for optimal image quality. Mounted slides may be stored at 4°C in the dark for several days, although best results are obtained when imaged within 1–2 days.

3.4 Imaging and Analysis

1. Use a fluorescence microscope equipped with appropriate filters. Identify cells at low magnification (10× or 20×) in the DAPI channel, then switch to a 63× or 100× oil-immersion objective. Apply immersion oil and adjust focus carefully.
2. Image using the green/orange filter set or a 561 nm laser (for confocal imaging). NEAT1 foci should appear as bright nuclear puncta. Adjust exposure time or laser power to prevent saturation. Verify that the cytoplasmic background remains low, with distinct nuclear signals.
3. Record DAPI for nuclei. If co-staining was performed (*see* **Note 7**), acquire those channels as well. Keep exposure consistent across conditions. Acquire z-stacks at 0.3–0.5 μm intervals to capture paraspeckles in different focal planes.
4. In mock-infected cells, ~5–10 paraspeckles per nucleus are typically observed as small (~0.5 μm) foci, often adjacent to SC35 speckles. HSV-1-infected cells (6–8 h) display increased numbers and intensities of NEAT1 foci, sometimes with slight enlargement (Fig. 2a). Record multiple nuclei per condition to ensure reproducibility.
5. Quantify paraspeckle number and intensity using ImageJ or equivalent software (*see* **Note 8**). Count foci per nucleus by thresholding the FISH channel, excluding small background particles. In blinded analyses of >20 cells, HSV-1–infected samples frequently exhibit ~2-fold increases in foci. Quantify total NEAT1 fluorescence per nucleus as an additional measure of expression, and apply appropriate statistical tests (e.g., Student's *t*-test) to compare conditions.

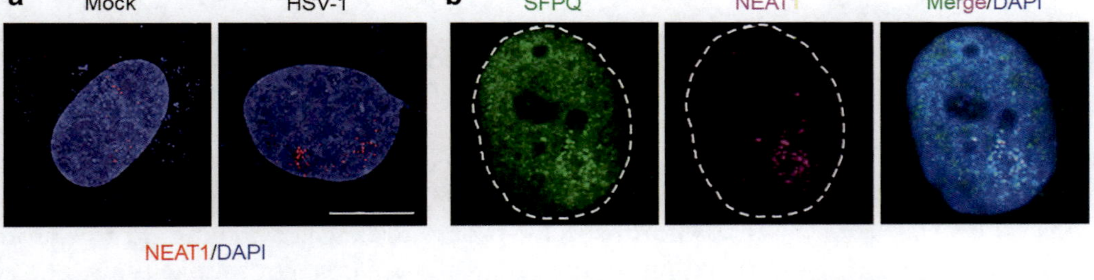

Fig. 2 NEAT1 RNA FISH in HSV-1-infected HeLa cells and co-localization with SFPQ. (**a**) HeLa cells were either mock-infected or infected with HSV-1 at an MOI of 5 for 4 h, followed by RNA fluorescence in situ hybridization (FISH) to detect NEAT1 lncRNA (red). Nuclei were counterstained with DAPI (blue). Increased NEAT1 puncta are observed in HSV-1–infected cells compared with mock controls. Scale bar, 10 μm. (**b**) HeLa cells subjected to NEAT1 RNA FISH (magenta) were subsequently analyzed by immunofluorescence staining for SFPQ (green) using Alexa Fluor 488. Nuclei were visualized with DAPI (blue). Merged images show co-localization of NEAT1 and SFPQ within paraspeckle structures

4 Notes

1. Cell density and health: Seeding density strongly affects infection outcome. Cells must be healthy and subconfluent to ensure uniform infection. Over-confluent or stressed cells show altered NEAT1 and fewer paraspeckles. Always include a mock-infected control to establish a baseline. HeLa cells constitutively express NEAT1, but overgrowth itself can increase levels. Plate cells to ~70% confluency at infection and ~90% by fixation. For other cell types, adjust accordingly.
2. Infection conditions: The timing and MOI should match the experimental goal. An MOI of 5–10 with HSV-1 yields high infection rates with synchronized entry. Short infections (4–6 h) activate immediate-early and early viral genes and can already induce NEAT1. Longer infections (>8 h) promote late gene expression and host shutoff, which may reduce NEAT1 if cells are stressed. Six to eight hours is a suitable starting point to observe paraspeckle responses before a strong cytopathic effect. Always include a mock-infected control, and when comparing viruses, equalize timing and conditions.
3. RNase-free technique: Before fixation, always use RNase-free reagents and tools (PBS, pipette tips, glassware). After fixation, RNA is crosslinked and more stable, but precautions reduce background problems. Gloves are mandatory. If RNase contamination is suspected (weak signals in both control and experiment), treat coverslips after ethanol permeabilization with 70% formamide/2× SSC at 37°C for 5 min to inactivate RNases and relax RNA secondary structure.
4. Probe selection (NEAT1 isoforms): Human NEAT1 has two isoforms, NEAT1_1 (~3.7 kb) and NEAT1_2 (~23 kb). Only NEAT1_2 supports paraspeckle assembly. Most probe sets target common regions, labeling both isoforms, but foci largely reflect NEAT1_2. For isoform-specific analysis, design probes to the unique 3′ region of NEAT1_2. Viral infections can alter NEAT1_2 processing, producing spliced or circular RNAs that cannot form paraspeckles. For such studies, combine general and isoform-specific probes to detect switching. In this protocol, common-region probes are used for simplicity.
5. Hybridization conditions: Standard conditions are overnight hybridization at 37°C in 10% formamide, which works well for most probes and cell types. To increase stringency, raise formamide to 20% or temperature. If the signal is weak, check probe integrity and increase probe concentration up to 250 nM, but avoid higher levels to prevent background. Do not agitate during hybridization; static incubation in a humid chamber is best. To prevent evaporation, Parafilm can be placed over the drop, but handle carefully when removing.

6. Image acquisition parameters: Use identical exposure or laser settings for all samples to allow comparison. NEAT1 foci are usually bright; set exposure to avoid saturation and merging. Confocal microscopy with Z-stacks (0.3–0.5 μm steps) improves resolution, and deconvolution can further separate close foci. Paraspeckles are near the diffraction limit (~0.5 μ m). Check for bleed-through between DAPI and Quasar570; if necessary, acquire sequentially.
7. Combining RNA FISH with immunofluorescence: RNA FISH can be combined with IF to visualize proteins such as paraspeckle factors (PSPC1, SFPQ) or viral proteins (ICP8, ICP27) (Fig. 2b). After FISH washes, fix again lightly (e.g., 2% PFA, 5 min), block, and incubate with primary antibody for 1 h, then secondary antibody for 45 min. Use fluorophores distinct from the RNA probe (e.g., Alexa488 if NEAT1 probe is in red). This approach shows co-localization of proteins with NEAT1 foci. Sequential FISH–IF is recommended for troubleshooting, though some protocols allow simultaneous staining.
8. Quantification and interpretation: Counting NEAT1 foci per nucleus provides a measure of paraspeckle assembly, but each focus may contain multiple transcripts, and large foci may represent fused paraspeckles. Measuring total NEAT1 intensity complements foci counts. Replicates often vary, so analyze multiple cells per condition. HSV-1 often increases foci from ~5 in control to ~15 in infected nuclei.
9. Safety considerations: Work under BSL-2 conditions. HSV-1 is an opportunistic pathogen. Wear gloves, a lab coat, and disinfect with bleach or 70% ethanol. Paraformaldehyde is toxic; handle it in a fume hood and dispose of it properly. Formamide is toxic and teratogenic; avoid inhalation and skin contact. Inactivate all viral waste (e.g., with bleach) before disposal.

Acknowledgments

This work was supported by the National Research Foundation of Korea (NRF), with grants from the Ministry of Science and ICT (RS-2024-00352590) and the Ministry of Education (RS-2023-00270936).

References

1. Spector DL, Lamond AI (2011) Nuclear speckles. Cold Spring Harb Perspect Biol 3(2):a000646. https://doi.org/10.1101/cshperspect.a000646
2. Li K, Wang Z (2021) Speckles and paraspeckles coordinate to regulate HSV-1 genes transcription. Commun Biol 4(1):1207. https://doi.org/10.1038/s42003-021-02742-6

3. Mao YS, Zhang B, Spector DL (2011) Biogenesis and function of nuclear bodies. Trends Genet 27(8):295–306. https://doi.org/10.1016/j.tig.2011.05.006
4. Duan Y et al (2023) LncRNA MALAT1 promotes growth and metastasis of head and neck squamous cell carcinoma by repressing VHL through a non-canonical function of EZH2. Cell Death Dis 14(2):149. https://doi.org/10.1038/s41419-023-05667-6
5. Hao M et al (2025) Paraspeckle component 1: a multifunctional RNA binding protein. Am J Cancer Res 15(5):2338–2352. https://doi.org/10.62347/riwh3000
6. Knutsen E, Harris AL, Perander M (2022) Expression and functions of long non-coding RNA NEAT1 and isoforms in breast cancer. Br J Cancer 126(4):551–561. https://doi.org/10.1038/s41416-021-01588-3
7. Major AT et al (2019) Dynamic paraspeckle component localisation during spermatogenesis. Reproduction 158(3):267–280. https://doi.org/10.1530/rep-19-0139
8. Bhattacharya A et al (2024) MUC1-C regulates NEAT1 lncRNA expression and paraspeckle formation in cancer progression. Oncogene 43(28):2199–2214. https://doi.org/10.1038/s41388-024-03068-3
9. Fox AH, Lamond AI (2010) Paraspeckles. Cold Spring Harb Perspect Biol 2(7):a000687. https://doi.org/10.1101/cshperspect.a000687
10. Hirose T et al (2014) NEAT1 long noncoding RNA regulates transcription via protein sequestration within subnuclear bodies. Mol Biol Cell 25(1):169–183. https://doi.org/10.1091/mbc.e13-09-0558
11. Milcamps R, Michiels T (2024) Involvement of paraspeckle components in viral infections. Nucleus 15(1):2350178. https://doi.org/10.1080/19491034.2024.2350178
12. Boros FA et al. (2025) Distinct expression of NEAT1 isoforms in Parkinson's disease models suggests different roles of the variants during the disease course. Sci Rep 15(1):13027. https://doi.org/10.1038/s41598-025-95787-0
13. Sciabica KS, Dai QJ, Sandri-Goldin RM (2003) ICP27 interacts with SRPK1 to mediate HSV splicing inhibition by altering SR protein phosphorylation. EMBO J 22(7):1608–1619. https://doi.org/10.1093/emboj/cdg166
14. Wang Z et al (2017) NEAT1 modulates herpes simplex virus-1 replication by regulating viral gene transcription. Cell Mol Life Sci 74(6):1117–1131. https://doi.org/10.1007/s00018-016-2398-4
15. Imamura K et al (2014) Long noncoding RNA NEAT1-dependent SFPQ relocation from promoter region to paraspeckle mediates IL8 expression upon immune stimuli. Mol Cell 53(3):393–406. https://doi.org/10.1016/j.molcel.2014.01.009
16. Landeras-Bueno S, Jorba N, Pérez-Cidoncha M, Ortín J (2011) The splicing factor proline-glutamine rich (SFPQ/PSF) is involved in influenza virus transcription. PLoS Pathog 7(11):e1002397. https://doi.org/10.1371/journal.ppat.1002397
17. Ma H et al (2017) The long noncoding RNA NEAT1 exerts antihantaviral effects by acting as positive feedback for RIG-I signaling. J Virol 91(9):e02250. https://doi.org/10.1128/jvi.02250-16
18. Zhang Q, Chen CY, Yedavalli VS, Jeang KT (2013) NEAT1 long noncoding RNA and paraspeckle bodies modulate HIV-1 posttranscriptional expression. mBio 4(1):e00596-12. https://doi.org/10.1128/mBio.00596-12
19. Qu D et al (2019) Long noncoding RNA MALAT1 releases epigenetic silencing of HIV-1 replication by displacing the polycomb repressive complex 2 from binding to the LTR promoter. Nucleic Acids Res 47(6):3013–3027. https://doi.org/10.1093/nar/gkz117
20. Harper KL et al (2024) Virus-modified paraspeckle-like condensates are hubs for viral RNA processing and their formation drives genomic instability. Nat Commun 15(1):10240. https://doi.org/10.1038/s41467-024-54592-5

Part III

Methods for the Functional Characterization of RNA-Proteins and RNA-RNA Interactions

Check for updates

Chapter 13

RBProximity-CLIP: A Method for Profiling the Interactome of an RNA-Binding Protein at Subcellular Resolution

Iwona Nowak, Hang T. Huynh, Mahekdeep Kaur, Orli Korchev, Veronica Moskovicz, Vivian Lobo, Michelle Fong, Aishe A. Sarshad and Daniel Benhalevy

Abstract

RNA-binding proteins (RBPs) are central players in post-transcriptional gene regulation (PTGR), and their mode of action often depends on subcellular localization. Conventional crosslinking and immunoprecipitation (CLIP) methods map the RNA targets of RBPs with nucleotide resolution but lack spatial specificity, making it impossible to distinguish how one RBP can contribute to biological processes across distinct compartments simultaneously. Here, we present *RBProximity-CLIP*, an extension of Proximity-CLIP that integrates APEX2-mediated proximity labeling with UV crosslinking of 4-thiouridine-labeled RNAs, followed by sequential biotin- and RBP-specific immunoprecipitations, to capture RBP–RNA complexes with both spatial and protein specificity. *RBProximity-CLIP* generates nucleotide-resolution maps of RBP–RNA interactions and enables proteomic characterization of complete ribonucleoprotein (RNP) compositions. Using Y-box binding protein 1 (YBX1) as a model RBP, we provide a detailed protocol for *RBProximity-CLIP*.

Keywords Crosslinking and immunoprecipitation (CLIP), RNA-binding proteins (RBPs), Protein–RNA interactions, Subcellular resolution, Spatial transcriptomics, YBX1, Multifunctional RBPs, Multilocalized RBPs, Post-transcriptional gene regulation, Proximity-biotinylation, APEX2

1 Introduction

Post-transcriptional gene regulation (PTGR) orchestrates all processes throughout the life cycle of mRNAs, including splicing, nuclear export, subcellular localization, translation, and degradation [1–3]. Central to PTGR are RNA-binding proteins (RBPs), which modulate and drive RNA metabolism [4–11]. RBPs are increasingly recognized as multifunctional regulators whose roles vary not only with

Iwona Nowak, Hang T. Huynh, and Mahekdeep Kaur contributed equally.

Orli Korchev and Veronica Moskovicz contributed equally and are listed by alphabetical order.

Je-Hyun Yoon (ed.), *RNA Labeling*, Methods in Molecular Biology, vol. 3047,
https://doi.org/10.1007/978-1-0716-5352-4_13,

their target RNA sequences but also with the cellular context, often manifesting through subcellular localization [12–14]. Thus, characterizing their functions within predefined cellular compartments is key to understanding how PTGR and RBPs shape gene expression.

Studying RBP–RNA interactions at subcellular resolution can provide insight into how cells execute RNA regulatory programs, as RBPs often display distinct functional roles depending on their localization [15–17]. A compelling example is Argonaute 2 (AGO2), which primarily functions in cytoplasmic miRNA-mediated gene silencing under normal conditions but relocates to the nucleus in response to both excentric and intrinsic signal cues. Examples include changes in cell density, viral infection, and oncogenetic settings [18–21], where it targets different RNA populations and affects cytoplasmic miRNA repression of its canonical targets [18–21]. Similarly, Y-box binding protein 1 (YBX1) exhibits context-dependent functions: while predominantly cytoplasmic under normal conditions, where it is implicated in regulating mRNA translation and stability, YBX1 translocates to the nucleus upon transcriptional stress or during specific cell cycle phases, where it functions as a transcriptional regulator [22, 23]. TAR DNA-binding protein 43 (TDP-43) provides another example: nuclear TDP-43 is retained through RNA-mediated macromolecular assembly and primarily regulates splicing, whereas cytoplasmic mislocalization, frequently observed in neurodegenerative diseases, results in altered RNA target specificity and pathological aggregation [24, 25].

Moreover, the same RBP can serve distinct biological functions depending on where it binds within target RNAs and its subcellular location. For instance, ELAV-like RNA-binding protein 1 (ELAVL1), also known as HuR, binds cytoplasmic mRNAs at AU-rich elements to stabilize them under normal conditions; however upon heat shock, it forms stress granules where it adopts an RNA-free state while simultaneously binding nuclear poly(A)+ RNA, suggesting distinct nuclear versus cytoplasmic RNA recognition modes [26, 27]. Another example is tRNA ligase, which performs conventional nuclear tRNA splicing under normal conditions but, upon ER stress, it acts near the ER membrane to finalize splicing of HAC1/XBP1 transcription factor mRNA in the regulation of the ER stress response [28, 29]. These spatially and functionally distinct roles highlight the critical need for methods that can detect RBP–RNA interactions with subcellular resolution [15, 30–37].

Understanding spatial PTGR is also critical because disruptions in RBP localization are implicated in a range of diseases. For example, in neurodegeneration, mislocalization of RBPs such as TDP-43 and FUS leads to aberrant RNA metabolism and toxic aggregation [38–47]. In cancer, rewiring of RNA regulatory networks through altered RBP localization supports malignant phenotypes [48, 49]. Hence, dissecting RBP–RNA interactions with subcellular precision is key to uncovering disease mechanisms and identifying novel therapeutic targets.

Traditional methods for studying RBP–RNA interactions, including crosslinking and immunoprecipitation (CLIP) approaches such as PhotoActivatable Ribonucleoside-enhanced CrossLinking and ImmunoPrecipitation (PAR-CLIP), have enabled identification of RNA targets of a given RBP at nucleotide resolution on a transcriptome-wide scale [50–56]. However, these approaches generally analyze whole-cell extracts, erasing the spatial context critical to RBP function. Other approaches, such as biochemical fractionation [57], can enrich for certain organelles, but most subcellular domains are unfractionable. In addition, cell fractionation is limited in capturing dynamic or transient localization and often suffers from contamination issues. Such limitations prevent a comprehensive understanding of subcellular RNA regulation.

To overcome these challenges, proximity labeling techniques, most notably those using the engineered soy ascorbate peroxidase, APEX2, fused to subcellular localization elements, have emerged as powerful tools for spatially resolved proteomics. APEX2 catalyzes the oxidation of biotin-phenol in the presence of H_2O_2 to generate biotin-phenoxyl radicals that biotinylate proteins within ~20 nm, enabling compartment-specific protein capture without fractionation [58–63]. Proximity-CLIP combines APEX2-based spatial labeling with UV crosslinking of RBPs to 4-thiouridine-labeled RNAs, enabling isolation of spatially defined ribonucleoprotein (RNP) complexes with improved specificity [30, 33, 36, 37, 64–66].

Here, we present a protocol for *RBProximity-CLIP*, which builds upon Proximity-CLIP by combining APEX2-mediated proximity biotinylation and protein–RNA crosslinking, followed by two sequential affinity purifications performed in tandem (Fig. 1). First, anti-biotin immunoprecipitation captures biotinylated, compartment-specific RNPs. Then, a second immunoprecipitation targets the specific RBP of interest, providing both RBP- and spatial-specificity. Notably, *RBProximity-CLIP* incoportates photoactivatable nucleotides and provides higher precision in RNP detection and size selection, offering 10- to 100-fold greater sensitivity, superior

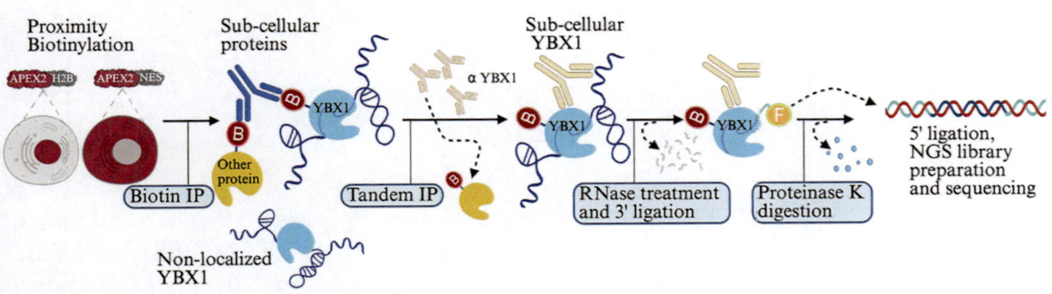

Fig. 1 Schematic representation of RBProximity-CLIP using YBX1 as a model protein. The workflow consists of two tandem immunoprecipitation steps: biotinylated protein enrichment followed by YBX1 pulldown. The procedure is illustrated in two distinct subcellular localizations, directed by APEX2-fused localization element (NES for cytoplasm and H2B for nucleus)

specificity, and improved versatility compared to colocalization-CLIP, an analogous method that was recently published [15]. Collectively, *RBProximity-CLIP* enables high-confidence mapping of RNA targets bound by individual RBPs in defined subcellular microenvironments.

We anticipate that the *RBProximity-CLIP* protocol will provide a unique and powerful window into the spatial complexity and multifunctionality of RBPs by bridging molecular specificity with cellular geography. By revealing compartment- and protein-specific RNA targets, this approach enables addressing previously inaccessible biological questions, identifying not only subcellular protein–RNA interactions but also entire RNP compositions. This capability opens new opportunities to study how PTGR dynamically responds to cellular signals and stress, advancing our mechanistic understanding of post-transcriptional regulation and how the spatial organization of RNA regulatory networks underpins cellular function in health and disease.

The following *RBProximity-CLIP* protocol will use YBX1 as a test case, resolving its nuclear and cytoplasmic subpopulations.

2 Materials

2.1 Generation of Stable Cell Lines Expressing Fused APEX2-Localization Elements

1. Flp-In™ T-REx™ HEK293 Cell Line (R78007, Invitrogen).
2. Flp-In™ T-REx™ Core Kit (K650001, Invitrogen).

2.2 Cell Growth, Proximity-Biotinylation, and Crosslinking

1. DMEM.
2. FBS.
3. Penicillin–Streptomycin (PenStrep).
4. Blasticidin.
5. Hygromycin B.
6. Poly-L-ornithine hydrobromide.
7. Doxycycline.
8. 4-Thiouridine (4SU).
9. Biotin tyramide (BP).
10. Hydrogen peroxide solution.
11. Quenching solution (freshly prepared): 10 mM sodium ascorbate (in PBS), 10 mM sodium azide (in PBS), and 5 mM Trolox (6-hydroxy-2,5,7,8-tetramethylchroman-2-carboxylic acid; in DMSO; 238813, Sigma-Aldrich).
12. Liquid nitrogen.

2.3 Cell Lysis

1. RIPA buffer: 50 mM Tris–HCl, 150 mM NaCl, 0.1% (wt/vol) SDS, 0.5% (wt/vol) sodium deoxycholate, 1% (wt/vol) Triton X-100. Adjust pH to 7.5 with HCl.
2. cOmplete™ EDTA-free Protease Inhibitor Cocktail (04693132001, Roche).
3. Sonicator.
4. Pierce™ 660 nm Protein Assay Reagent (10177723, Fisher).
5. Bovine serum albumin (BSA).

2.4 Immunoprecipitation of Biotin-Labeled Proteins

1. Invitrogen™ Dynabeads™ Protein G for Immunoprecipitation (10003D, Invitrogen).
2. Anti-biotin antibody (ab53494, Abcam).
3. Acetic acid.
4. Sodium hydroxide.
5. Universal pH paper strips 1–14 (1.10962, MilliporeSigma).
6. Alfa Aesar™ Laemmli SDS Sample Buffer, reducing, 6× (15493939, Fisher).
7. PageRuler™ Plus Prestained Protein Ladder, 10 to 250 kDa (26619, Thermo Scientific).
8. SuperSignal™ West Dura Extended Duration Substrate (10220294, Fisher).
9. SDS-PAGE 4%–12%, Bis-Tris, 1.0 mm Mini Protein Gel, 12-well (15324604, Fisher).
10. Cytiva Amersham™ Protran™ Premium NC nitrocellulose membrane (15269794, Cytiva).
11. ChemiDoc MP Imaging Systems (12003154, Biorad).

2.5 Immunoprecipitation of Subcellular YBX1 from the Biotin-Labeled Fraction

Anti-YBX1 antibody (ab12148, Abcam).

2.6 Preparation of Small RNA cDNA Libraries from RNP Footprints for NGS

2.6.1 Reagents

1. RNase T1 (1000 U/μl) (EN0541, Thermo Scientific).
2. Quick CIP (M0525, NEB).
3. RNase Inhibitor (037, A&A Biotechnology).
4. T4 RNA Ligase 2, truncated K227Q (M0351, NEB).
5. T4 Polynucleotide Kinase (M0201, NEB).

6. 10 mM ATP.
7. Proteinase K, recombinant, PCR Grade (RPROTK-RO, Roche).
8. Phenol–chloroform, water saturated.
9. Chloroform, water saturated.
10. Isopropanol.
11. Ethanol.
12. GlycoBlue™ Coprecipitant (15 mg/mL) (AM9515, Invitrogen).
13. T4 RNA Ligase 1 (EL0021, Thermo Scientific).
14. Oligo Clean & Concentrator Kit (D4060, Zymo Research).
15. Deoxynucleotides (dNTPs) Solution Mix.
16. SuperScript™ IV Reverse Transcriptase (18090050, Invitrogen).
17. Platinum™ Taq DNA Polymerase (10966018, Invitrogen).
18. Gel Breaker Tubes (Fisher Scientific)/0.5 mL tube needle-punctured at the bottom several times.
19. Invitrogen™ UltraPure™ Low Melting Point (LMP) Agarose (16520050, Invitrogen).
20. 3% Agarose, dye-free, w/internal standards, Pippin Prep, 100–250 bp, 10/pk, Marker P (CDP3010, Bionordika).

2.6.2 Buffers

1. RNase T1 buffer: 20 mM Tris, pH 7.4, 150 mM NaCl, 2 mM EDTA, 1% NP40.
2. Dephosphorylation buffer: 50 mM Tris–HCl, pH 7.5, 100 mM NaCl, 10 mM $MgCl_2$.
3. PNK buffer: 50 mM Tris–HCl, pH 7.5, 50 mM NaCl, 10 mM $MgCl_2$.
4. Proteinase K buffer: 50 mM Tris, pH 7.5, 75 mM NaCl, 6.25 mM EDTA, 1% SDS.

2.6.3 Oligos

1. Fluorescent 3′ adapter: /5rApp/NNTGACTGTGGAATTCTCGGGT/iAlex660N/GCCAAGG/3AlexF660N/.
2. 5′ adapter: 5′-(aminolinker)-GTTCAGAGTTCTACAGTCCGACGATCrNrNrNrN.
3. RT primer: 5′-GCCTTGGCACCCGAGAATTCCA.
4. 5′ short primer: CTTCAGAGTTCTACAGTCCGACGA.
5. 5′ long primer: AATGATACGGCGACCACCGAGATCTACACGTTCAGAGTTCTACAGTCCGA.

6. 3′ RNA index primer: CAAGCAGAAGACGGCATACGAGATGCAATTCGGTGACTGGAGTTCCTTGGCACCCGAGAATTCCA.

3 Methods

3.1 Generation of Stable Cell Lines Expressing Fused APEX2-Localization Element

A stable Flp-In™ T-REx™ HEK293 cell line expressing APEX2 linked to a localization element of interest (NES and H2B in this protocol) was generated using the Flp-In™ T-REx™ System, according to the manufacturer's instructions. **See Notes 1 and 2** about cell line generarion, pay special attention to **Note 3** about cell line validation, and Figure 2a in [30].

3.2 Cell Growth, Proximity-Biotinylation, and Crosslinking

1. Coat cell culture plates with 0.01% poly-L-ornithine hydrobromide for 2 h at 37°C. Aspirate the poly-L-ornithine hydrobromide solution and rinse once with 1× PBS.
2. Seed Flp-In™ T-REx™ 293 APEX2-localization element expressing cells at 35,000 cells/cm^2 in DMEM supplemented with 10% FBS and 1:100 v/v PenStrep (*see* **Note 4**).
3. After 24 h, add doxycycline (final concentration: 2 μg/mL) and 4SU (final concentration: 100 μM) to the culture. This supplementation step should be performed no more than 16 h before continuing with the protocol (*see* **Note 5**).
4. The next day, label the plate both on the lids and sides. Dissolve the biotin-phenol (BP) in DMSO to a stock solution of 500 mM. Add BP to all plates, except no-BP controls, to achieve a final working concentration of 500 μM. If needed, dilute BP in pre-warmed medium before adding to the cells. Gently swirl and return plates to the incubator for 30 min (*see* **Note 6**).
5. During the incubation time, prepare the quenching solution and aliquot it into 50 mL tubes (*see* **Note 7**).
6. After incubation, add 100 mM hydrogen peroxide to the plates to achieve a final concentration of 1 mM. Gently swirl and incubate exactly 1 min at room temperature. *Do not exceed the 1-min incubation time.*
7. Discard the culture medium by decanting.
8. Immediately wash cells three times with quenching solution. For rapid quenching, pour the solution directly from 50 mL tubes. Use at least 1 mL per 3 cm^2 surface area.
   Important: You need to ensure beforehand that your cell lines already inducibly express the APEX2 constructs and can bioti-

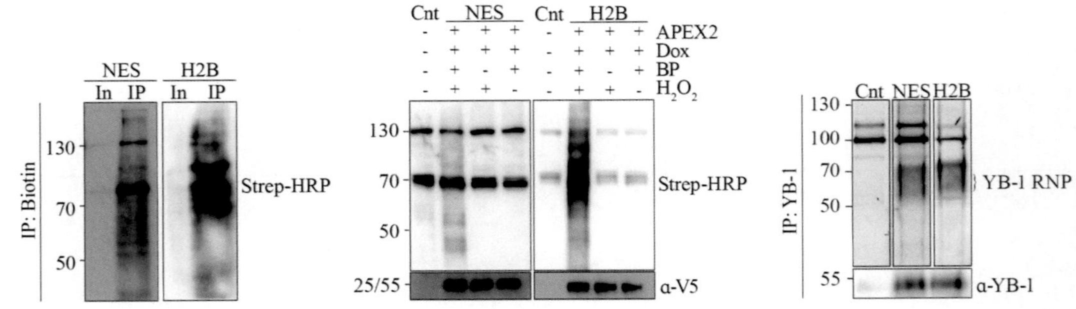

Fig. 2 Validation of APEX2-dependent protein biotinylation. Representative immunoblots of biotinylated proteins and V5-tagged APEX2 constructs from whole-cell lysates of Flp-In™ T-REx™ 293 cells expressing V5-APEX2-NES (~25 kDa) or V5-APEX2-H2B (~55 kDa). Cells were treated with biotin-phenol (BP) and hydrogen peroxide (H_2O_2) to induce APEX2-mediated protein biotinylation, or with incomplete reaction components (without BP or H_2O_2) as negative controls. Parental cells were included as an additional negative control (Cnt)

nylate nearby proteins in the presence of BP and hydrogen peroxide (see Fig. 2 in this chapter and Figure 2a in [30]).

9. After the third wash, decant the quenching solution (do not aspirate the quenching solution; leaving a small volume on the cells facilitates their collection after UV crosslinking). Remove the lids and immediately place the plates in a UV crosslinker. Irradiate the cells with 312 nm UV light to deliver 0.15 J/cm^2 for 5 min.
10. After crosslinking, place the plates on ice.
11. Using a cell scraper, detach the cells and transfer the cell suspension in leftover quenching solution into the collection tubes (*see* **Note 8**).
12. Pellet the cells by centrifugation at 300 × *g* for 5 min at 4°C. Carefully aspirate and discard the supernatant.
13. Snap-freeze the cell pellets in liquid nitrogen and store at –80°C or proceed directly to cell lysis.

3.3 Cell Lysis

1. Supplement ice-cold RIPA buffer with 10 mM sodium azide, fresh 10 mM sodium ascorbate, fresh 5 mM Trolox, and 1× protease inhibitor cocktail.
2. Add the supplemented RIPA buffer to the cell pellets (300 μL for 10 cm plates and 900 μL for 15 cm plates). Resuspend the pellets by pipetting up and down a few times. Incubate lysates on ice for 30 min. The cell suspension can also be sonicated for 30 s on and 30 s off three times.
3. Clear extracts by centrifugation at 21,000 × *g* for 15 min at 4°C. Transfer supernatant to a labeled, pre-chilled tube without

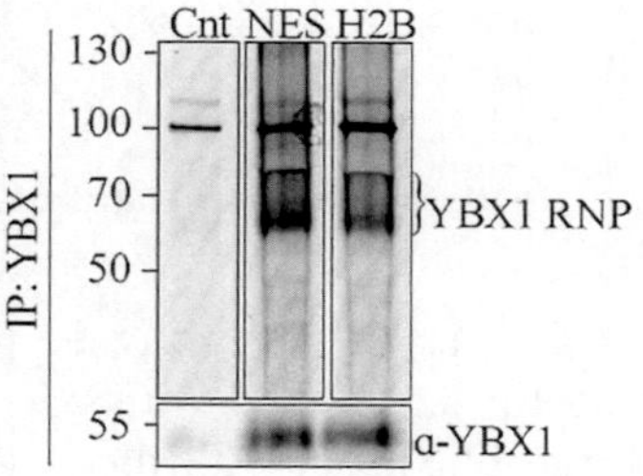

Fig. 3 Immunoprecipitation (IP) of biotinylated proteins. Representative immunoblots showing IP of biotinylated proteins from whole-cell lysates of Flp-In™ T-REx™ 293 cells expressing V5-APEX2-NES (NES) or V5-APEX2-H2B (H2B). Biotinylated proteins were immunoprecipitated using an anti-biotin antibody bound to Protein G magnetic beads. A constant amount of cell extract (In, 10 μg) was loaded in the first lane, and the corresponding immunoprecipitated fraction is shown in the second lane (IP, 5%)

disturbing the debris. Keep 2%–5% of the input at −80°C for later analysis (*see* **Note 9** and Fig. 3).

4. Quantify protein concentration using the Pierce™ 660 nm Protein Assay, following the manufacturer's instructions. Bradford or BCA reagents are not compatible.

3.4 Immunoprecipitation of Biotin-Labeled Proteins

1. Wash 25 μL Protein G magnetic beads three times with RIPA buffer.
2. Bind the anti-biotin antibody to pre-washed beads by incubating for 4 h at 4°C with rotation or shaking for continuous mixing of beads and antibody (*see* **Note 10**).
3. Wash antibody-bound magnetic beads three times with 1 mL RIPA buffer to remove unbound antibody.
4. Add clarified cell extract from Sect. 3.3, step 3 to antibody-bound magnetic beads, ensuring a total protein input of 0.5–3 mg per reaction. Incubate overnight at 4°C with gentle end-to-end rotation.
5. The following day, briefly spin down any beads from the top of the tube and place the tubes on a magnetic rack to collect beads. Transfer the flow-through to a clean tube for later analysis (store at −80°C until use). Wash the beads three times with 1 mL RIPA buffer.
6. Elute biotinylated proteins from beads by adding 150 μL of 2% (v/v) acetic acid and gently pipetting to resuspend. Transfer eluate to a pre-labeled 1.5 mL microcentrifuge tube using a magnetic rack.
7. Neutralize eluates immediately by adding 1 M sodium hydroxide dropwise while gently mixing, until pH reaches 6–8. Use pH paper strips, but do not insert paper into the tube; instead, at each measurement point, pipette 1 μL of the eluate onto the

paper. Save 10 μL of the eluate for later analysis (store at –80°C until use) (*see* Fig. 3).

3.5 Immunoprecipitation of Subcellular YBX1 from the Biotin-Labeled Fraction

1. Wash 25 μL Protein G magnetic beads three times with RIPA buffer.
2. Bind anti-YBX1 antibody to the pre-washed beads by incubating for 4 h at 4°C with rotation or shaking for continuous mixing of beads and antibody (*see* **Note 10**).
3. Magnetize the beads, discard flow-through, and wash three times with 1 mL RIPA buffer to make sure only well-bound antibody remains.
4. Add the neutralized eluates (from Sect. 3.4, step 7) from anti-biotin IP to magnetic beads. Incubate overnight at 4°C with gentle rotation.
5. Briefly spin down any beads from the top of the tube and place the tubes on a magnetic rack to collect beads. Transfer the flow-through to a clean tube for later analysis (store at –80°C until use). Wash beads three times with 1 mL RIPA buffer. Proceed with washed beads to small RNA library preparation for NGS.
6. Please *see* **Note 11** for an alterative method of performing the tandem IP.

3.6 Preparation of Small RNA cDNA Libraries from RNP Footprints for NGS

3.6.1 On-Beads RNase Treatment

1. Resuspend beads in 50 μL of 1 U/μL RNase T1 prepared in RNase T1 buffer. Incubate for 15 min at 22°C with shaking at 800 rpm. Make sure no RNase remnants reach external parts of the tube, tube racks, or gloves. *It is recommended to change gloves after pipetting the RNase into the reaction mix.*
2. Wash beads three times with 1 mL RIPA buffer and once with 1 mL dephosphorylation buffer. Make sure no RNase carryover on the side or lid of tubes.

3.6.2 On-Beads Dephosphorylation

1. Prepare the following reaction master mix.

Dephosphorylation master mix	
Component	**×1 RXN (μL)**
10× CIP buffer (*cutsmart*)	3
5 U/μl Quick CIP	3
40 U/μl RNase Inhibitor	1

Dephosphorylation master mix	
Component	**×1 RXN (μL)**
RNase-free ddH_2O	23
Total reaction volume	30

2. Incubate beads with 30 μL reaction master mix for 10 min at 37°C with shaking at 800 rpm.
3. Wash beads twice with 1 mL dephosphorylation buffer and twice with 1 mL PNK buffer.

3.6.3 On-Beads 3′ Adapter Ligation

1. Prepare the following 3′ adapter ligation master mix:

3′ Adapter ligation master mix	
Component	**×1 RXN (μL)**
50 μM Fluorescent 3′ adapter	0.5
10× RNA ligase buffer (NEB)	5
50% aqueous PEG-8000	15
T4 RNA Ligase 2, truncated K227Q	2.5
40 U/μl RNase Inhibitor	1
RNase-free ddH_2O	26
Total reaction volume	50

2. Resuspend beads in 50 μL of 3′ adapter ligation master mix.
3. Incubate overnight at 4°C with gentle agitation to prevent beads from settling (*see* **Note 12**).
4. Wash beads four times with 1 mL PNK buffer.

3.6.4 On-Beads Phosphorylation

1. Prepare the following PNK reaction master mix:

Phosphorylation master mix	
Component	**×1 RXN (μL)**
10× PNK buffer (NEB)	5
40 U/μl RNase Inhibitor	1
T4 Polynucleotide Kinase	1
10 mM ATP (cold)	5
RNase-free ddH_2O	38
Total reaction volume	50

2. Resuspend beads in 50 μL PNK reaction master mix. Incubate for 30 min at 37°C with shaking at 800 rpm.
3. Wash beads three times with 1 mL of RIPA buffer.

3.6.5 SDS-PAGE Purification

1. Place tubes on a magnetic rack to collect the beads. Discard the supernatant.
2. Resuspend beads in 40 μL of 3× sample buffer, vortex vigorously, and briefly centrifuge to collect the suspension.
3. Elute RNPs by heating at 95°C for 5 min. Briefly centrifuge to collect the sample.
4. Place tubes on a magnetic rack and transfer the supernatant to a clean, pre-labeled 1.5 mL microcentrifuge tube. *Optional stopping point*: store samples at −80°C until further processing.
5. Load 36 μL of each sample on an SDS-PAGE, leaving one well empty between samples to prevent cross-contamination. In the empty wells, load 5 μL of a 1:10 dilution of PageRuler Plus molecular weight marker (in ddH_2O) to facilitate the identification and excision of the band corresponding to YBX1 in Sect. 3.6.5, step 7. Retain the remaining 4 μL of each sample for western blot analysis (see Fig. 4 lower panel).
6. Upon completion of electrophoresis, place the gel inside a clear plastic sheet protector and scan for IR680 fluorescence to detect the fluorophore signal. *Add a 100 μL ddH_2O on top of the gel to prevent drying.*
7. Carefully excise the gel band corresponding to the RBP plus the 3′-ligated adapter, located approximately 25 kDa above the free protein's expected position. Use prior western blot results to determine the migration position of your protein (not the theoretical size) and add ~25 kDa to locate the correct band (*see* **Note 13** and Fig. 4 upper panel).

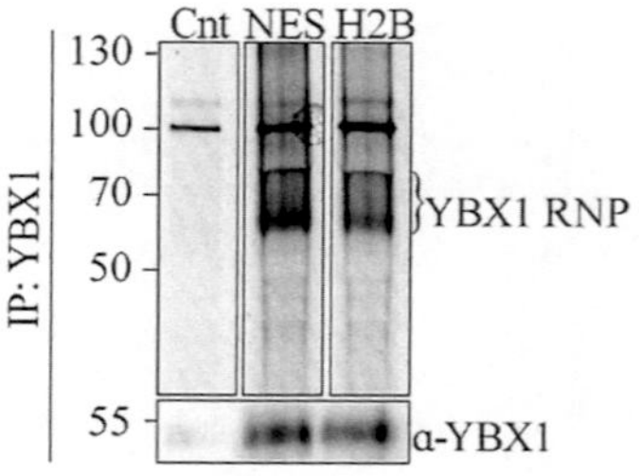

Fig. 4 Examination of subcellular YBX1 RNP complexes. Fluorescently imaged SDS-PAGE of 3′-ligated crosslinked YBX1 RNP complexes (upper panel) and a parallel immunoblot of YBX1 (lower panel) following anti-YBX1 IP in biotinylated Flp-In™ T-REx™ 293 cells expressing V5-APEX2-NES (NES) or V5-APEX2-H2B (H2B). Parental cells were included as a negative control (Cnt)

8. Transfer the gel slice to a gel breaker tube and centrifuge at 21,000 × *g* for 1 min at 4°C. Repeat if gel fragments remain in the breaker tube.

3.6.6 Proteinase K Treatment

1. Boil the shredded gel in 100 μL 2× Proteinase K buffer supplemented with 50 mM DTT for 2 min at 95°C.
2. Release the RNA from the shredded gel by Proteinase K digestion in three consecutive steps, adding freshly prepared Proteinase K solution incrementally to the existing volume each time. Pre-weigh Proteinase K into three separate tubes and add buffer immediately before use to maintain freshness:
   2.1 Add 400 μL of 2.4 mg/mL Proteinase K solution in Proteinase K buffer. Incubate at 50°C in a thermoblock with shaking at 1000 rpm for 30 min.
   2.2 Add 250 μL of 1.5 mg/mL Proteinase K solution in Proteinase K buffer. Incubate at 50°C in a thermoblock with shaking at 1000 rpm for 30 min.
   2.3 Add 250 μL of 1.5 mg/mL Proteinase K solution in Proteinase K buffer. Incubate at 50°C in a thermoblock with shaking at 1000 rpm for 30 min.
3. Centrifuge at 21,000 × *g* for 5 min. Carefully transfer the supernatant into a clean 2 mL microcentrifuge tube, avoid disturbance of the gel pellet. Repeat if necessary to avoid any gel carryover.

3.6.7 RNA Isolation

Note: All steps involving chloroform and phenol–chloroform must be performed in a chemical hood.

1. Add 1 mL phenol–chloroform to the 1 mL sample and mix by vortexing.
2. Centrifuge at 12,000 × *g* for 10 min at 4°C.
3. Transfer the upper aqueous phase to a new 2 mL microcentrifuge tube.
4. Add 1 mL chloroform and mix by vortexing.
5. Centrifuge at 12,000 × *g* for 2 min at 4°C.
6. Transfer the upper aqueous phase to a new 2 mL microcentrifuge tube.
7. Precipitate the RNA by adding 1.5 μL GlycoBlue and 1 mL isopropanol. Mix by vortexing and incubate at −80°C for at least 1 h. *Optional stopping point*: samples may be stored at −80°C overnight.
8. Centrifuge at 21,000 × *g* for 20 min at 4°C. Carefully remove all supernatant without disturbing the visible blue pellet.

9. Wash the pellet by gently adding 1 mL of ice-cold 75%–80% ethanol without resuspending the pellet. Centrifuge at 21,000 × *g* for 5 min. Carefully remove all traces of ethanol. If residual ethanol remains, perform a series of short centrifugation spins and carefully aspirate the remaining liquid using a pipette.
10. Air-dry the pellet for 5 min at room temperature.
11. Dissolve the pellet in 8.5 μL of RNase-free ddH_2O.

3.6.8 5′ Adapter Ligation

1. Prepare the following 5' adapter mastermix:

5′ Adapter ligation master mix	
Component	**×1 RXN (μL)**
50 μM 5′ adapter	0.2
10× RNA ligase buffer with ATP	2
50% PEG-8000	6
RNase-free ddH_2O	0.3
Total reaction volume	8.5

2. Add 8.5 μL of 5′ adapter ligation master mix to each sample. Incubate at 90°C for 1 min to denature the RNA (*see* **Note 14**).
3. Cool the reaction on ice for 2 min.
4. Add 2 μL T4 RNA Ligase 1 and 1 μL RNase Inhibitor. Mix gently and incubate at 37°C for 1 h.
5. Add 30 μL RNase-free ddH_2O to reach a final volume of 50 μL.
6. Purify the reaction product using the Oligo Clean & Concentrator Kit following the manufacturer's instructions. Perform the final elution in 14 μL RNase-free ddH_2O.

3.6.9 Reverse Transcription (RT)

1. Anneal the RT primer to the template RNA by adding 12 μL of the following mix to each sample:
   Ncubate at 65°C for 5 min.

Component	**×1 RXN (μL)**
100 μM RT Primer	2
dNTP mix (2mM each)	10
Total reaction volume	12

2. Chill the reaction on ice for at least 1 min.
3. Add 14 μL reverse transcription reaction master mix to each sample:

Reverse transcription master mix	
Component	**×1 RXN (μL)**
5× SSIV buffer	8
100 mM DTT	2
40 U/μl RNase Inhibitor	2
Superscript IV	2
Total reaction volume	14

4. Mix well, spin down, and incubate at 56°C for 30 min.
5. Inactivate the reaction by incubating at 80°C for 10 min.
6. Add 60 μL RNase-free ddH_2O to the reverse transcription reaction to reach a final cDNA volume of 100 μL.

3.6.10 Low Cycle PCR with Short Primers

1. Prepare the following master mix for amplification of the cDNA by PCR:

Low cycle PCR master mix	
Component	**×1 RXN (μL)**
10× Platinum PCR buffer	40
dNTP mix (2 mM)	10
50 mM $MgCl_2$	12
100 μM 5′ short primer	0.8
100 μM RT primer	0.8
Platinum™ Taq DNA Polymerase	1.6
RT product (cDNA)	40
RNase-free ddH_2O	294.8
Total reaction volume	400

2. Divide the reaction into eight PCR tubes with 50 μL each. Run a seven-cycle PCR with the following program:

Low cycle PCR plan		
# of cycles	**Temperature**	**Time**
1	94°C	2 min
7	94°C	30 s
	60°C	30 s
	72°C	15 s
1	72°C	1 min

3. Combine the PCR tubes into a 2 mL microcentrifuge tube. Concentrate the sample using the Oligo Clean & Concentrator Kit following the manufacturer's instructions. Elute in 30 µL of RNase-free ddH_2O.
4. Perform size selection of the amplified product (75–100 nt) on a 3% agarose Pippin Prep cassette according to the manufacturer's instructions. The final eluate volume should be 40 µL. If less is recovered, adjust by adding RNase-free ddH_2O to reach 40 µL.

3.6.11 Pilot PCR

1. To determine the optimal PCR cycle number for sufficient, yet not excessive, amplification of the indexed library, perform a pilot PCR with the following reaction mix:
   See **Note 15** regarding 3' RNA Index Primer

Pilot PCR master mix	
Component	**×1 RXN (µL)**
10× Platinum PCR buffer	10
dNTP mix (2 mM each)	2.5
50 mM $MgCl_2$	3
100 µM 5' long primer	0.2
100 µM 3' RNA index primer	0.2
Platinum™ Taq DNA Polymerase	0.4
RT product (cDNA)	10
RNase-free ddH_2O	73.7
Total reaction volume	100

2. Divide the reaction into seven PCR tubes with 14 µL each. Run the following thermocycler program:

Pilot PCR plan		
# of cycles	**Temperature**	**Time**
1	94°C	2 min
27	94°C	30 s
	60°C	30 s
	72°C	15 s
1	72°C	1 min

3. Remove one tube after each of these cycle numbers: 9, 12, 15, 18, 21, 24, and 27.

4. Run the PCR products at 100 V on a 2.5% low melting point agarose gel. Identify the optimal cycle number as the lowest cycle where the desired product (approx. 140–180 bp) becomes visible. Empty library (ligated 3' and 5' adapter with no insert) appears near 126 bp.

3.6.12 Preparative PCR Amplification

1. Use the same reagents as in the pilot PCR to set up an identical reaction but with a volume of 300 µL. Divide equally into four PCR tubes (75 µL each) and run the thermocycler program using the previously determined optimal PCR cycle number.
2. Combine the PCR products, then clean and concentrate them using the Oligo Clean & Concentrator Kit following the manufacturer's instructions. Elute in 30 µL RNase-free ddH_2O.
3. Perform another size selection of 140–180 nt fragments using a 3% agarose Pippin Prep cassette, following the manufacturer's instructions. The final eluate volume should be 40 µL.
4. Assess the quality and quantity of the library using TapeStation analysis before submitting for NGS.

4 Notes

1. **Generation of Stable Cell Lines Expressing Fused APEX2-Localization Elements:**
   APEX2 Function and Mechanism APEX2 is an engineered soybean ascorbate peroxidase that enables spatially controlled protein labeling within cells [61]. In the presence of hydrogen peroxide, APEX2 catalyzes the oxidation of biotin-phenol (BP), producing highly reactive biotin-phenoxyl radicals with a half-life of less than 1 ms. While most of these short-lived radicals are quenched by water, a fraction reacts covalently with nearby proteins at aromatic amino acid residues, resulting in their biotinylation. By directing APEX2 to specific subcellular compartments through the fusion with a localization element (in this case, NES and H2B for cytoplasmic and nuclear localization, respectively), proteins in defined cellular environments can be selectively labeled.
2. **Generation of Stable Cell Lines, Cell Culture, and Adaptability of the Protocol**
   In this protocol, stable cell lines expressing the V5-APEX2-NES/H2B fusion proteins were generated using the Flp-In™ T-REx™ system, following the manufacturer's protocol. This system allows for controlled expression of the APEX2 constructs, ensuring regulated experimental conditions. However, it is important to note that APEX2 fusion proteins can also

be introduced into cells through alternative methods, such as lentiviral transduction or transient transfection, depending on experimental needs.

Stable cell lines expressing the V5-APEX2-NES/H2B fusion proteins were cultured in Dulbecco's Modified Eagle Medium (DMEM) supplemented with 10% (v/v) fetal bovine serum (FBS), 1:100 v/v PenStrep, 100 μg/mL hygromycin B, and 5 μg/mL blasticidin. Please note that when seeding cells for the experiment, the DMEM medium does not need to contain 100 μg/mL hygromycin B and 5 μg/mL blasticidin. Moreover, the principles and steps outlined in this protocol are adaptable to a wide range of cell lines, including those that grow in suspension rather than adherent monolayers.

3. **Validation of APEX2-Dependent Protein Biotinylation**
Biochemical controls of proximity biotinylation are critical to ensure specificity and reliability of the labeling process. Western blotting with HRP-coupled streptavidin (Strep-HRP) on cell extracts confirms that biotinylation occurs only when all key components: APEX2, biotin-phenol (BP), and H_2O_2 are present. This verifies the reaction specificity and rules out nonspecific background labeling. After preparing the cell line, perform proximity-biotinylation and cell lysis to validate APEX2-dependent protein biotinylation in the cell line before proceeding to the *RBProximity-CLIP* experiment. See Fig. 2 for an example of how the Western blot should appear. Furthermore, immunofluorescence can be performed to validate localized active APEX2 in the cell line (see Figure 2a in [30] as an example).

4. **Cell Growth, Proximity-Biotinylation, and Crosslinking**:
Cell Seeding
The optimal cell seeding density should be adjusted individually for each cell line. It is important to keep a balance between achieving a sufficiently high number of cells for downstream applications with maintaining the cells' capacity for rapid proliferation, which is essential for the efficient incorporation of 4SU into newly synthesized RNA. In our protocol, we aimed for the cells to reach approx. 90% confluency within 48 h post-seeding. In addition, the size and number of culture plates required for each experiment should be calibrated based on the abundance of the protein of interest within the cells and the efficiency of the IP. For proteins expressed at low levels, larger culture areas or multiple plates may be necessary to obtain sufficient material. Specifically, for Flp-In™ T-REx™ HEK293 cells, note that they detach easily from cell culture plates, so handling must be gentle to avoid losing cells. Do not pour liquids directly onto the cells; instead, gently add liquids to the plate via the sides. For multiple solution

exchanges (e.g., washes), it is recommended to avoid pipetting and instead pour from pre-measured 50 mL tubes.

5. **Induction of APEX2-Localization Element Expression**
V5-APEX2-NES/H2B expression in our system is controlled by a Tet-ON system. However, doxycycline, an analog of tetracycline, is used instead, due to its higher stability and avidity.

6. **Preparation and Handling of Biotin-Phenol (BP)**
The biotin-phenol stock solution should be prepared in dimethyl sulfoxide (DMSO) to ensure solubility and stability. We recommend preparing a 500 mM stock solution of BP in DMSO, which can then be aliquoted in small volumes (e.g., 500 μL) and stored at −80°C to prevent degradation over time. To fully dissolve the biotin phenol powder in DMSO, sonication may be necessary. Proper dissolving is critical for consistent labeling efficiency during the APEX2 reaction.
To perform the BP reaction in live cells, remove the existing culture medium and replace it with pre-warmed DMEM supplemented with 500 μM BP. For a 10-cm dish, prepare this by diluting 10 μL of a 500 mM BP stock solution in 10 mL of pre-warmed DMEM before adding it to the dish. The BP stock should never be pipetted directly onto the cells; always dilute it into medium prior to cell treatment to ensure complete solubilization.

7. **Preparation of Quenching Solution**
The quenching solution must be freshly prepared before each experiment to effectively halt the APEX2-catalyzed biotinylation reaction. Sodium azide can be prepared in advance as a stable 1 M stock solution in RNase-free ddH_2O, which can be stored in aliquots at −20°C for later use. In contrast, sodium ascorbate and Trolox are unstable and should be freshly prepared on the day of the experiment. Sodium ascorbate is diluted in RNase-free ddH_2O, while Trolox is dissolved in DMSO. The final quenching solution is prepared in fresh PBS by combining these components immediately before use.

8. **Cell Collection**
After the quenching steps, a small volume of quenching solution typically remains in the culture plates, which can be used to help detach and collect the cells. If this volume is insufficient, add up to 750 μL of fresh quenching solution to facilitate the cell collection. Avoid adding larger volumes. Use a cell scraper to gently detach the cells, taking care to preserve pellet integrity. Transfer the cell suspension into the tubes for downstream processing.

9. **Immunoprecipitation Steps:***Recommended Controls*
Biochemical controls of proximity biotinylation are critical to ensure specificity and reliability of the labeling process.

Western blotting with HRP-coupled streptavidin on both cell extracts and eluates confirms that biotinylation occurs only when all key components—APEX2, biotin-phenol (BP), and hydrogen peroxide—are present. This verifies the reaction specificity and rules out nonspecific background labeling. Moreover, analyzing the unbound fractions ("flow-through") alongside the eluates helps confirm that the biotinylated proteins were efficiently captured by the beads. This step ensures minimal loss of target proteins and demonstrates enrichment of the protein of interest.

10. **Bead-to-Antibody and Cell Extract Ratio Calibration**
    Calibrating the optimal ratio between the bead volume and the antibody, as well as between the antibody-bound magnetic beads and cell extract, is essential for each APEX2-fusion protein cell line and antibody used. Antibody-to-beads ratio calibration applies to both immunoprecipitation steps in this protocol, biotin IP and YBX1 IP, and directly impacts the efficiency and specificity of protein capture. Using too few beads risks losing biotinylated or target proteins that remain unbound in the solution, reducing the yield and sensitivity of the experiment. Conversely, excess beads can increase nonspecific binding, raising background noise and complicating data interpretation.
    Monitoring the unbound fraction (flow-through) after incubation provides a reliable indication of protein capture efficiency. If a significant amount of biotinylated/target protein remains in the flow-through, it suggests insufficient bead volume or antibody concentration, indicating the need for further optimization. Based on our experience and antibodies, 10 μg of anti-biotin antibody and 3–4 μg of anti-YBX1 antibody per 25 μL of Protein G beads have proven effective. However, these amounts may need adjustment depending on antibody affinity and sample specifics.

11. **Alternative Way of Preforming the IP**
    Immunoprecipitation can also be performed by adding specific (anti-YBX1 or anti-biotin) antibody to cell lysates and incubating overnight at 4°C with gentle rotation. The following day, 25 μL of pre-washed Protein G magnetic beads are added and incubated for 2 h at room temperature with gentle rotation to capture the antibody–protein complexes. Beads are then washed three times with RIPA buffer to remove nonspecific interactions.

12. **Preparation of Small RNA cDNA Libraries from RNP Footprints for NGS:** *Fluorescent 3′ Adapter*
    The fluorescent 3′ adapter has the following structure:
    5′-rAppNNTGACTGTGGAATTCTCGGGT(fl)GCCAAGG-fl

- "rApp" indicates a 5′ RNA adenylated phosphate group required for ligation.
- "NN" represents a randomized (wild cards) two-base sequence, which provides a unique molecular identifier (UMI) and prevents sequence-based ligation bias.
- The underlined sequence (index) consists of any known six-base combination that serves as a sample-specific barcode.
- The "(fl)" denotes the fluorescent label, which can be one of the following dye pairs: 2×ATTO647N, 2×ATTO665, 2× Alexa Fluor 647 (AF647), or 2×Alexa Fluor 660 (AF660). In this protocol, *AF660 was used.*

13. **SDS-PAGE Band Excision**
    Place the gel inside a clear plastic sheet protector and image it. Use the size marker loaded in-between each sample to identify the correct protein size. Cut the band using a sterile scalpel and put the tube inside a tub shredder. Take a new image of the gel after you have excited the band to ensure you cut the correct bands.
14. **5′ Adapter**
    The 5′ index primer used in this protocol has the following sequence:
    5′-(aminolinker)-GTTCAGAGTTCTACAGTCCGACGAT-CrNrNrNrN
    - A 5′ amine group
    - "NNNN" represents a randomized (wild cards) four-base sequence, which provides a unique molecular identifier (UMI) and prevents sequence-based ligation bias.
15. **3′ Index Primer**
    The 3′ index primer used in this protocol has the following sequence:
    CAAGCAGAAGACGGCATACGAGATGCAATTCGGT-GACTGGAGTTCCTTGGCACCCGAGAATTCCA

This primer corresponds to the P74-G2 index from the NEB-Next® Multiplex Oligos for Illumina® (96 Index Primers) kit, which uses eight-base indexes for multiplexing. To multiplex multiple samples for sequencing, different 3′ index primers, each with a unique index sequence (underlined), can be assigned to label each sample distinctly. This allows to pool several samples together for efficient sequencing and later de-multiplex the reads based on these indexes.

References

1. Shyu A, Wilkinson MF, van Hoof A (2008) Messenger RNA regulation: to translate or to degrade. EMBO J 27:471–481. https://doi.org/10.1038/sj.emboj.7601977
2. Müller-McNicoll M, Neugebauer KM (2013) How cells get the message: dynamic assembly and function of mRNA–protein complexes. Nat Rev Genet 14:275–287. https://doi.org/10.1038/nrg3434
3. Keene JD (2007) RNA regulons: coordination of post-transcriptional events. Nat Rev Genet 8:533–543. https://doi.org/10.1038/nrg2111
4. Hentze MW, Castello A, Schwarzl T, Preiss T (2018) A brave new world of RNA-binding proteins. Nat Rev Mol Cell Biol 19:327–341. https://doi.org/10.1038/nrm.2017.130
5. Gerstberger S, Hafner M, Tuschl T (2014) A census of human RNA-binding proteins. Nat Rev Genet 15:829–845. https://doi.org/10.1038/nrg3813
6. Zhang S, Yang X, Jiang M, et al (2022) Post-transcriptional control by RNA-binding proteins in diabetes and its related complications. Front Physiol 13. https://doi.org/10.3389/fphys.2022.953880
7. Glisovic T, Bachorik JL, Yong J, Dreyfuss G (2008) RNA-binding proteins and post-transcriptional gene regulation. FEBS Lett 582:1977–1986. https://doi.org/10.1016/j.febslet.2008.03.004
8. Harvey RF, Smith TS, Mulroney T et al. (2018) Trans-acting translational regulatory RNA binding proteins. Wiley Interdiscip Rev: RNA 9:e1465. https://doi.org/10.1002/wrna.1465
9. Sangokoya C, Blelloch R (2024) Coordinate post-transcriptional regulation by microRNAs and RNA binding proteins is critical for early embryonic cell fate decisions. bioRxiv. 2024.10.02.616359. 10.1101/2024.10.02.616359
10. Marondedze C (2020) The increasing diversity and complexity of the RNA-binding protein repertoire in plants. Proc R Soc B 287:20201397. https://doi.org/10.1098/rspb.2020.1397
11. Fierro-Monti I (2024) RBPs: an RNA editor's choice. Front Mol Biosci 11:1454241. https://doi.org/10.3389/fmolb.2024.1454241
12. Schneider-Lunitz V, Ruiz-Orera J, Hubner N, van Heesch S (2021) Multifunctional RNA-binding proteins influence mRNA abundance and translational efficiency of distinct sets of target genes. PLoS Comput Biol 17:e1009658. https://doi.org/10.1371/journal.pcbi.1009658
13. Van Nostrand EL, Freese P, Pratt GA et al. (2020) A large-scale binding and functional map of human RNA-binding proteins. Nature 583:711–719. https://doi.org/10.1038/s41586-020-2077-3
14. Grifone R, Shao M, Saquet A, Shi D-L (2020) RNA-binding protein Rbm24 as a multifaceted post-transcriptional regulator of embryonic lineage differentiation and cellular homeostasis. Cells 9:1891. https://doi.org/10.3390/cells9081891
15. Yi S, Singh SS, Rozen-Gagnon K, Luna JM (2024) Mapping RNA–protein interactions with subcellular resolution using colocalization CLIP. RNA 30:920–937. https://doi.org/10.1261/rna.079890.123
16. Khoroshkin M, Buyan A, Dodel M et al. (2024) Systematic identification of post-transcriptional regulatory modules. Nat Commun 15: 7872. https://doi.org/10.1038/s41467-024-52215-7
17. Kejiou NS, Palazzo AF (2017) mRNA localization as a rheostat to regulate subcellular gene expression. WIREs RNA 8:e1416. https://doi.org/10.1002/wrna.1416
18. Johnson KC, Kilikevicius A, Hofman C et al. (2024) Nuclear localization of Argonaute 2 is affected by cell density and may relieve repression by microRNAs. Nucleic Acids Res 52:1930–1952. https://doi.org/10.1093/nar/gkad1155
19. Lobo V, Shcherbinina E, Westholm JO et al. (2024) Integrative transcriptomic and proteomic profiling of the effects of cell confluency on gene expression. Sci Data 11: 617. https://doi.org/10.1038/s41597-024-03465-z
20. Lobo V, Nowak I, Fernandez C et al. (2024) Loss of Lamin A leads to the nuclear translocation of AGO2 and compromised RNA interference. Nucleic Acids Res 52:9917–9935. https://doi.org/10.1093/nar/gkae589
21. Huang H-C, Fong M, Nowak I et al. (2025) Nuclear AGO2 supports influenza A virus replication through type-I interferon regulation. Nucleic Acids Res 53:gkaf268. https://doi.org/10.1093/nar/gkaf268
22. Cohen SB, Ma W, Valova VA et al. (2010) Genotoxic stress-induced nuclear localization of oncoprotein YB-1 in the absence of proteolytic processing. Oncogene 29:403–410. https://doi.org/10.1038/onc.2009.321
23. Sogorina EM, Kim ER, Sorokin AV et al. (2021) YB-1 phosphorylation at serine 209 inhibits its nuclear translocation. Int J Mol Sci 23:428. https://doi.org/10.3390/ijms23010428
24. dos Passos PM, Hemamali EH, Mamede LD et al. (2024) RNA-mediated ribonucleoprotein

assembly controls TDP-43 nuclear retention. PLoS Biol 22:e3002527. https://doi.org/10.1371/journal.pbio.3002527
25. Duan L, Zaepfel BL, Aksenova V, et al (2022) Nuclear RNA binding regulates TDP-43 nuclear localization and passive nuclear export. Cell Rep 40. https://doi.org/10.1016/j.celrep.2022.111106
26. Gallouzi I-E, Brennan CM, Stenberg MG et al. (2000) HuR binding to cytoplasmic mRNA is perturbed by heat shock. Proc Natl Acad Sci 97:3073–3078. https://doi.org/10.1073/pnas.97.7.3073
27. Peng SS, Chen CY, Xu N, Shyu AB (1998) RNA stabilization by the AU-rich element binding protein, HuR, an ELAV protein. EMBO J 17:3461. https://doi.org/10.1093/emboj/17.12.3461
28. Sidrauski C, Cox JS, Walter P (1996) tRNA ligase is required for regulated mRNA splicing in the unfolded protein response. Cell 87:405–413. https://doi.org/10.1016/S0092-8674(00)81361-6
29. Jurkin J, Henkel T, Nielsen AF et al. (2014) The mammalian tRNA ligase complex mediates splicing of XBP1 mRNA and controls antibody secretion in plasma cells. EMBO J 33:2922–2936. https://doi.org/10.15252/embj.201490332
30. Benhalevy D, Anastasakis DG, Hafner M (2018) Proximity-CLIP provides a snapshot of protein-occupied RNA elements in subcellular compartments. Nat Methods 15:1074–1082.https://doi.org/10.1038/s41592-018-0220-y
31. Fazal FM, Han S, Parker KR et al. (2019) Atlas of subcellular RNA localization revealed by APEX-seq. Cell 178:473–490.e26. https://doi.org/10.1016/j.cell.2019.05.027
32. Kaewsapsak P, Shechner DM, Mallard W et al. (2017) Live-cell mapping of organelle-associated RNAs via proximity biotinylation combined with protein-RNA crosslinking. eLife 6:e29224. https://doi.org/10.7554/eLife.29224
33. Koch L (2019) Proximity-CLIP — close encounters of the RNA kind. Nat Rev Genet 20:68–69. https://doi.org/10.1038/s41576-018-0086-y
34. Li Y, Aggarwal MB, Ke K et al. (2018) Improved analysis of RNA localization by spatially restricted oxidation of RNA-protein complexes. Biochemistry 57:1577–1581. https://doi.org/10.1021/acs.biochem.8b00053
35. Padrón A, Iwasaki S, Ingolia NT (2019) Proximity RNA labeling by APEX-seq reveals the organization of translation initiation complexes and repressive RNA granules. Mol Cell 75:875–887.e5. https://doi.org/10.1016/j.molcel.2019.07.030
36. Anastasakis D, Benhalevy D, Hafner M (2020) Proximity-CLIP and expedited non-radioactive library preparation of small RNA footprints for next generation sequencing. Curr Protoc Mol Biol 131:e120. https://doi.org/10.1002/cpmb.120
37. Benhalevy D, Hafner M (2020) Proximity-CLIP provides a snapshot of protein-occupied RNA elements at subcellular resolution and transcriptome-wide scale. In: Heinlein M (Ed.) Eds. RNA tagging: methods and protocols. Springer US, New York, NY, 283–305
38. Farris S, Hacisuleyman E, Donlin-Asp P, Cioni J-M (2022) Editorial: RNA localization and localized translation in neurons. Front Integr Neurosci 15. https://doi.org/10.3389/fnint.2021.831038
39. Li Y, Sun S (2025) RNA dysregulation in neurodegenerative diseases. EMBO J 44:613–638. https://doi.org/10.1038/s44318-024-00352-6
40. Rhine K, Li R, Kopalle HM et al. (2025) Neuronal aging causes mislocalization of splicing proteins and unchecked cellular stress. Nat Neurosci 28:1174–1184. https://doi.org/10.1038/s41593-025-01952-z
41. Acosta-Galeana I, Hernández-Martínez R, Reyes-Cruz T et al (2023) RNA-binding proteins as a common ground for neurodegeneration and inflammation in amyotrophic lateral sclerosis and multiple sclerosis. Front Mol Neurosci 16. https://doi.org/10.3389/fnmol.2023.1193636
42. Jo M, Lee S, Jeon Y-M et al. (2020) The role of TDP-43 propagation in neurodegenerative diseases: integrating insights from clinical and experimental studies. Exp Mol Med 52:1652–1662. https://doi.org/10.1038/s12276-020-00513-7
43. Loganathan S, Lehmkuhl EM, Eck RJ, Zarnescu DC (2020) To be or not to be... toxic—is RNA association with TDP-43 complexes deleterious or protective in neurodegeneration? Front Mol Biosci 6. https://doi.org/10.3389/fmolb.2019.00154
44. Masaki K, Sonobe Y, Ghadge G et al. (2020) RNA-binding protein altered expression and mislocalization in MS. Neurol Neuroimmunol Neuroinflamm 7:e704. https://doi.org/10.1212/NXI.0000000000000704
45. Naumann M, Pal A, Goswami A et al. (2018) Impaired DNA damage response signaling by FUS-NLS mutations leads to neurodegeneration and FUS aggregate formation. Nat Commun 9: 335. https://doi.org/10.1038/s41467-017-02299-1
46. Harley J, Hagemann C, Serio A, Patani R (2021) TDP-43 and FUS mislocalization in VCP mutant motor neurons is reversed by

pharmacological inhibition of the VCP D2 ATPase domain. Brain Commun 3:fcab166. https://doi.org/10.1093/braincomms/fcab166

47. Daigle JG, Lanson NA, Smith RB et al. (2013) RNA-binding ability of FUS regulates neurodegeneration, cytoplasmic mislocalization and incorporation into stress granules associated with FUS carrying ALS-linked mutations. Hum Mol Genet 22:1193–1205. https://doi.org/10.1093/hmg/dds526
48. Mohibi S, Chen X, Zhang J (2019) Cancer the'RBP'eutics – RNA-binding proteins as therapeutic targets for cancer. Pharmacol Ther 203:107390. https://doi.org/10.1016/j.pharmthera.2019.07.001
49. Kang D, Lee Y, Lee J-S (2020) RNA-binding proteins in cancer: functional and therapeutic perspectives. Cancers (Basel) 12:2699. https://doi.org/10.3390/cancers12092699
50. Benhalevy D, McFarland HL, Sarshad AA, Hafner M (2017) PAR-CLIP and streamlined small RNA cDNA library preparation protocol for the identification of RNA binding protein target sites. Methods 118–119:41–49. https://doi.org/10.1016/j.ymeth.2016.11.009
51. Van Nostrand EL, Pratt GA, Shishkin AA et al. (2016) Robust transcriptome-wide discovery of RNA-binding protein binding sites with enhanced CLIP (eCLIP). Nat Methods 13:508–514. https://doi.org/10.1038/nmeth.3810
52. Licatalosi DD, Mele A, Fak JJ et al. (2008) HITS-CLIP yields genome-wide insights into brain alternative RNA processing. Nature 456:464–469. https://doi.org/10.1038/nature07488
53. Schneider C, Kudla G, Wlotzka W et al. (2012) Transcriptome-wide analysis of exosome targets. Mol Cell 48:422–433. https://doi.org/10.1016/j.molcel.2012.08.013
54. Hendrickson DG, Kelley DR, Tenen D et al. (2016) Widespread RNA binding by chromatin-associated proteins. Genome Biol 17:28. https://doi.org/10.1186/s13059-016-0878-3
55. Hafner M, Landthaler M, Burger L et al. (2010) Transcriptome-wide identification of RNA-binding protein and microRNA target sites by PAR-CLIP. Cell 141:129–141. https://doi.org/10.1016/j.cell.2010.03.009
56. Anastasakis DG, Jacob A, Konstantinidou P et al. (2021) A non-radioactive, improved PAR-CLIP and small RNA cDNA library preparation protocol. Nucleic Acids Res 49:e45. https://doi.org/10.1093/nar/gkab011
57. Huynh HT, Shcherbinina E, Huang H-C et al. (2024) Biochemical separation of cytoplasmic and nuclear fraction for downstream molecular analysis. Curr Protoc 4:e1042. https://doi.org/10.1002/cpz1.1042
58. Branon TC, Bosch JA, Sanchez AD et al. (2018) Efficient proximity labeling in living cells and organisms with TurboID. Nat Biotechnol 36:880–887. https://doi.org/10.1038/nbt.4201
59. de Boer E, Rodriguez P, Bonte E et al. (2003) Efficient biotinylation and single-step purification of tagged transcription factors in mammalian cells and transgenic mice. Proc Natl Acad Sci USA 100:7480–7485. https://doi.org/10.1073/pnas.1332608100
60. Fernández-Suárez M, Chen TS, Ting AY (2008) Protein-protein interaction detection *in vitro* and in cells by proximity biotinylation. J Am Chem Soc 130:9251–9253. https://doi.org/10.1021/ja801445p
61. Hung V, Udeshi ND, Lam SS et al. (2016) Spatially resolved proteomic mapping in living cells with the engineered peroxidase APEX2. Nat Protoc 11:456–475. https://doi.org/10.1038/nprot.2016.018
62. O'callaghan CA, Byford MF, Wyer JR et al. (1999) BirA enzyme: production and application in the study of membrane receptor-ligand interactions by site-specific biotinylation. Anal Biochem 266:9–15. https://doi.org/10.1006/abio.1998.2930
63. Yang J, Jaramillo A, Shi R et al. (2004) *In vivo* biotinylation of the major histocompatibility complex (MHC) class II/peptide complex by coexpression of BirA enzyme for the generation of MHC class II/tetramers. Hum Immunol 65:692–699. https://doi.org/10.1016/j.humimm.2004.04.001
64. Hafner M, Katsantoni M, Köster T et al. (2021) CLIP and complementary methods. Nat Rev Methods Primers 1:20. https://doi.org/10.1038/s43586-021-00018-1
65. Bonislawski (2018) Protein proximity labeling approach enables more precise investigation of RNA localization. In: GenomeWeb. https://www.genomeweb.com/proteomics-protein-research/protein-proximity-labeling-approach-enables-more-precise-investigation. Accessed 19 Oct 2025
66. Daniel Benhalevy, Markus Hafner, Dimitrios Anastasakis 2018. Proximity-CLIP.Protocol exchange https://dx.doi.org/10.1038/protex.2018.115

Chapter 14

RAP-RNA: A Methodological Approach to Study Base-Pairing Between Endogenous RNAs

Sylvia Wood, Bridget Flowers and Jong-Sun Lee

Abstract

Long non-coding RNAs (lncRNAs) engage in RNA-RNA interactions (RRIs) that shape transcript fate and gene regulation, yet these duplexes are often transient and difficult to capture in cells. This chapter presents RNA Antisense Purification-RNA coupled to psoralen crosslinking (RAP-RNA[AMT]), a targeted method to capture direct RNA-RNA base-paired partners of a chosen lncRNA in its native cellular context. Duplexes are stabilized with 4′-aminomethyltrioxsalen (AMT) and UVA; the bait lncRNA is captured with biotinylated antisense probes under high stringency, and RNA partners associated through direct base pairing are released by RNase H digestion and quantified by RT-qPCR. The protocol provides practical guidance on probe design, crosslinking parameters, essential controls, and quantitative analysis. Together, the step-by-step workflow enables high-specificity mapping of lncRNA-centered RRIs and can be adapted for focused validation across diverse targets and cell types.

Keywords RAP-RNA, AMT (psoralen) crosslinking, LncRNAs, RNA–RNA interactions, Antisense oligonucleotide capture

1 Introduction

RNA–RNA interactions (RRIs) are fundamental to gene expression, shaping transcript structure, localization, stability, and ultimately cell fate [1–3]. These interactions arise either within a single RNA, where intramolecular base pairing organizes secondary and higher-order structures, or between distinct RNAs, where pairing mediates molecular recognition and complex assembly [4]. RRIs are especially central to non-coding RNAs, whose functions often rely on base-paired recognition, scaffolding, and targeting during processes such as splicing, RNA modification, translation control, and decay [5–8]. Canonical examples include small nuclear RNAs (snRNAs) pairing with precursor messenger RNAs (pre-mRNAs) during splicing [9], small nucleolar RNAs (snoRNAs) guiding site-specific modification of ribosomal RNAs (rRNAs) [10], and microRNAs pairing with mRNAs to control translation and stability [11, 12]. Disruptions to

Je-Hyun Yoon (ed.), *RNA Labeling*, Methods in Molecular Biology, vol. 3047,
https://doi.org/10.1007/978-1-0716-5352-4_14,

these RRI networks are implicated in cancers, neurodegeneration, and congenital disorders, highlighting the importance of mapping RNA-RNA pairings to deepen our understanding of diseases [13–15].

Long non-coding RNAs (lncRNAs) engage diverse RNA-binding partners. Typically, longer than 200 nucleotides, they act in cis to modulate nearby chromatin and transcription, and in trans to pair with distal RNAs and scaffold ribonucleoprotein (RNP) assemblies [16, 17]. Their expression is often tissue-specific, and their subcellular localization is frequently cell-type and compartment-specific; many lncRNAs also show disease-associated expression patterns, supporting therapeutic target and biomarker development [18–22]. Despite these features and widespread links to disease, only a small fraction of the many thousands of annotated lncRNA loci have been mechanistically characterized [23]. To move from expression correlations to mechanism, functional annotation requires systematic profiling of the RRIs that mediate recognition, scaffolding, and targeting. This is challenging in cells: interactions are often transient or dynamic [24, 25], cue-dependent [26], and embedded within large RNP complexes [27], hindering confident detection and assignment [28]. These constraints motivate crosslinking-and-capture strategies that stabilize native duplexes prior to selective enrichments, thereby improving recovery of context-specific pairings [29–35].

Initial studies confirmed direct pairwise binding in vitro using low-throughput biochemical and biophysical assays, such as electrophoretic mobility shift assays (EMSA) [36–38] and single-molecule fluorescence resonance energy transfer (smFRET) [39, 40], but offered limited cellular context. Chemical probing approaches, including selective 2′-hydroxyl acylation analyzed by primer extension (SHAPE) [41, 42] and dimethyl sulfate mapping (DMS) [43, 44], together with computational modeling [45], provide broader structural coverage, yet they cannot unambiguously assign which RNAs pair with which across the transcriptome in cells [3]. A major inflection point came with psoralen-based crosslinking, which enables direct capture of RNA duplexes in cells [29–35]. Upon ultraviolet-A (UVA) illumination, the psoralen derivative 4′-aminomethyltrioxsalen (AMT) forms reversible inter-strand crosslinks between opposing pyrimidines (enriched at uridines) via cycloaddition at the C5=C6 double bond, transiently stabilizing native helices for purification and identification [46] (Fig. 1). Transcriptome-wide implementations such as Psoralen Analysis of RNA Interactions and Structures (PARIS) [29, 30], Ligation of Interacting RNA and high-throughput sequencing (LIGR-seq) [31], Sequencing of Psoralen-crosslinked, Ligated, And Selected Hybrids (SPLASH) [32], and Crosslinking Of Matched RNAs And Deep Sequencing (COMRADES) [33], combine psoralen crosslinking with proximity ligation and high-throughput sequencing, reading duplexes as chimeric junctions to produce global maps of intra- and intermolecular RNA contacts. While powerful for discovery at scale, these approaches prioritize

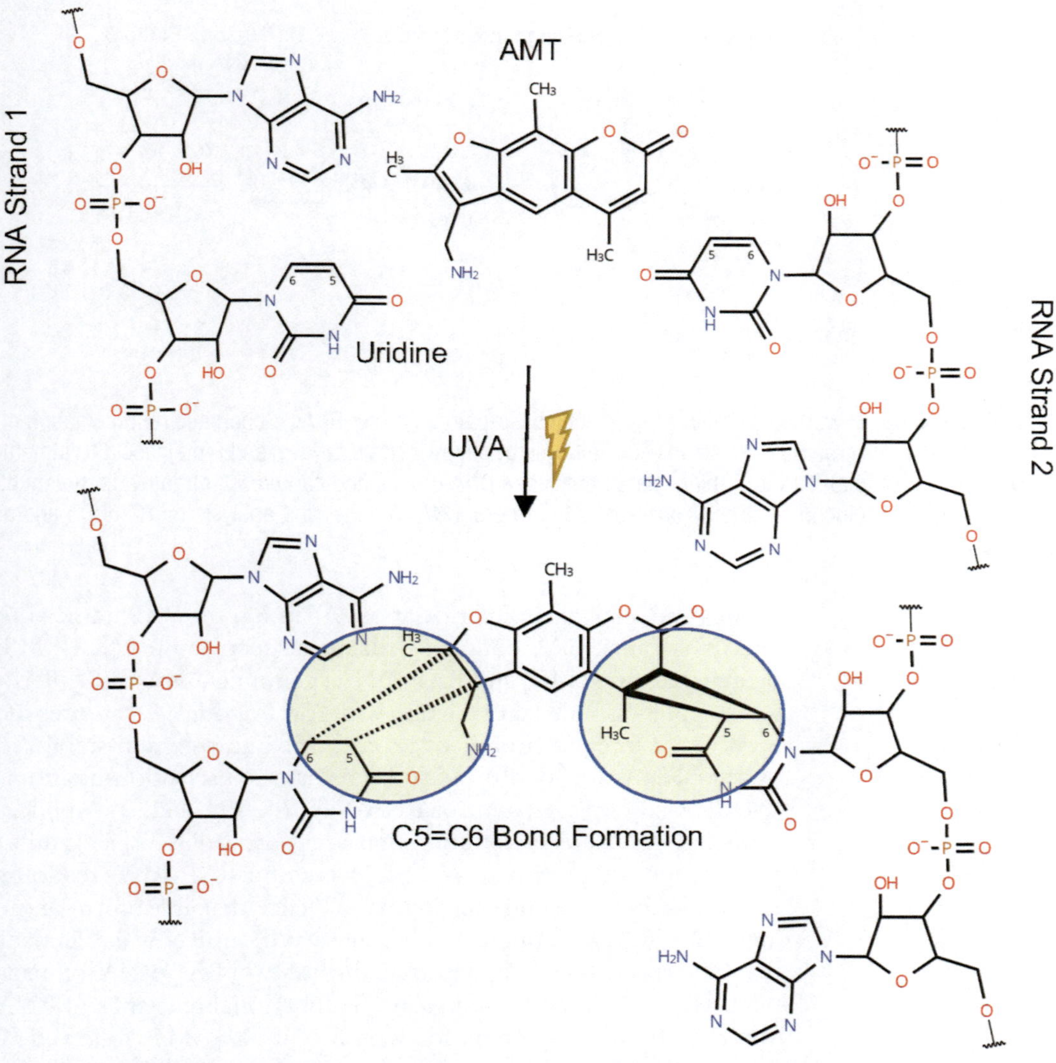

Fig. 1 AMT-mediated inter-strand crosslinking of opposing uridines. AMT intercalates between paired RNAs; UVA (~350 nm) induces [2 + 2] photocycloaddition at the uridine C5=C6 double bond to form a monoadduct, followed by a second reaction on the facing uridine to yield a covalent interstrand crosslink (highlighted). The adducts are photoreversible. All chemical structures used for this figure were drawn using the open-source chemical structure editor Ketcher (EPAM Systems, Version v2.25.0)

breadth over target specificity and, because chimeric junctions yield scales with transcript abundance, are comparatively insensitive for the exhaustive recovery of interactions around a single low-abundance lncRNA [29–33].

RNA Antisense Purification-RNA (RAP-RNA) offers a targeted, highly specific strategy to identify endogenous RNA duplexes [34]. Rather than globally selecting crosslinked helices, RAP-RNA first isolates a chosen lncRNA with long, tiled, biotinylated antisense DNA

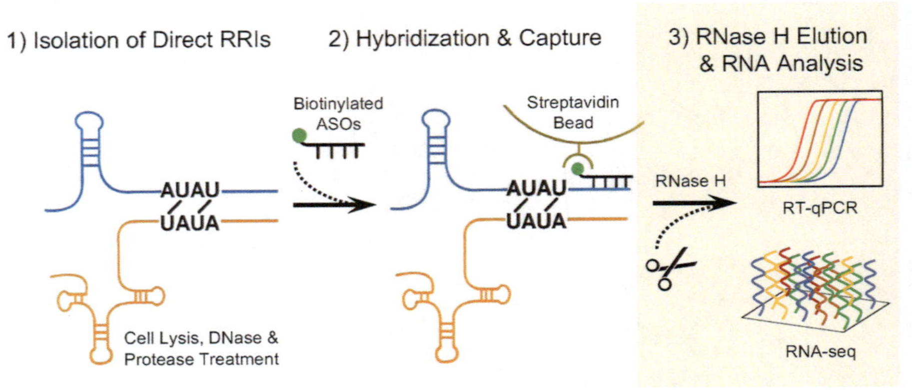

Fig. 2 RAP-RNA[AMT] workflow for isolating directly base-paired partner RNAs. Schematic representation of: (1) In-cell AMT + UVA crosslinks RNA duplexes, followed by chemical lysis, DNase, and Proteinase K treatment; (2) Hybridization of total RNA to 5′-biotinylated antisense DNA probes and capture on streptavidin magnetic beads; (3) RNase H elution of directly base-paired partners, RNA cleanup, and analysis by RT-qPCR and/or RNA-seq

probes under stringent hybridization. Adapting the RAP framework and its variants, RAP coupled with mass spectrometry (RAP-MS; protein partners) [47] and RAP-DNA (genomic DNA) [48], RAP-RNA applies psoralen crosslinking with AMT and UVA, captures the lncRNA of interest, and recovers co-purifying, psoralen-stabilized partner RNAs for identification by reverse transcription quantitative PCR (RT-qPCR) and/or RNA-seq [34, 49] (Fig. 2). Applications to U1 snRNA recognition of nascent pre-mRNAs [34] and to nuclear speckle-proximal splicing leveraging RAP-RNA readouts [50], together with complementary evidence that an isoform of the lncRNA *CRNDE* contains an ultraconserved, snoRNA-like element that binds pre-rRNA [49], illustrate the utility of RAP-RNA for mapping RRIs in cells and underscore the functional relevance of RNA duplexes. In this chapter, we focus on RAP-RNA[AMT] followed by RT-qPCR and provide a practical, step-by-step workflow covering experimental design, crosslinking parameters, probe tiling, capture, recovery, and analysis, with recommended controls and troubleshooting guidance to enable confident mapping and validation of base-paired RNA binding partners of endogenous RNAs. Collectively, this workflow affords high-specificity, quantitative validation of lncRNA-centered RRIs by RT-qPCR and is readily adapted to different targets, cell types, and perturbations.

2 Materials

2.1 786-O Cell Culture and Harvesting

1. 786-O cells (male).
2. RPMI 1640 Medium, HEPES (Invitrogen) supplemented with 10% fetal bovine serum (Sigma-Aldrich) and 1× Antibiotic-Antimycotic (Gibco).

3. Trypsin-EDTA (Gibco).
4. 1× Phosphate-buffered saline (PBS).

2.2 AMT Crosslinking

1. 4′-aminomethyltrioxsalen hydrochloride (Sigma).
2. 2× PBS.
3. Spectrolinker XL-1500 (Spectronics).
4. 0.3 mm plain glass plate to cover samples during irradiation.

2.3 Cell Lysis and RNA Extraction

1. Cell Lysis Buffer components (RNase-free): 10 mM Tris–HCl pH 7.5, 500 mM LiCl, 0.5% Triton X-100, 0.2% SDS, 0.1% sodium deoxycholate.
2. SUPERase·In RNase inhibitor (Thermo Fisher).
3. TURBO DNase (Thermo Fisher).
4. Proteinase K (NEB).
5. RNeasy Mini Kit (Qiagen).

2.4 Probe Hybridization, Capture, and Wash

1. LiCl Hybridization Buffer components (RNase-free): 10 mM Tris–HCl pH 7.5, 1 mM EDTA, 500 mM LiCl, 1% Triton X-100, 0.2% SDS, 0.1% sodium deoxycholate, 4 M urea.
2. B&W buffer: 10 mM Tris pH 7.5, 0.5 mM EDTA, 1 M NaCl, 0.1% tween 20.
3. Dynabeads MyOne Streptavidin C1 magnetic beads (Thermo Fisher).
4. Low Stringency Wash Buffer: 1× SSPE, 0.1% SDS, 1% NP-40, 4 M urea.
5. High Stringency Wash Buffer: 0.1× SSPE, 0.1% SDS, 1% NP-40, 4 M urea.

2.5 RNase H Elution and RNA Clean-Up

1. RNase H (NEB).
2. RNase H Elution Buffer: 20 mM Tris–HCl pH 7.5, 40 mM KCl, 8 mM $MgCl_2$, 1 mM DTT, 0.125% N-lauroylsarcosine, 0.025% sodium deoxycholate.
3. TRIzol (Invitrogen).
4. RNA Clean & Concentrator-5 columns (Zymo Research).

2.6 Reverse Transcription and qPCR

1. PrimeScript RT Master Mix (TaKaRa).
2. SYBR Green PCR Master Mix (Applied Biosystems).
3. Target-specific qPCR primers (70–150 bp amplicons; primer sites flanking but not overlapping the probe footprint).
4. Nuclease-free water, low-bind PCR plates, optical seals.
5. qPCR instrument and associated software.

3 Methods

3.1 Prep, Design, and Controls

1. Probe design (single-probe capture for RT-qPCR readout). Use a 40-60 nucleotide (nt) 5′-biotinylated antisense DNA oligonucleotide (ASO) complementary to the target, with ~40%–60% GC (35%–65% acceptable). If the target region is AU-rich, the probe may be extended up to ~70 nt to meet DNA:RNA Tm ≥65°C. Avoid polymorphic bases and highly structured or GC-clamped regions; screen for self-dimers/hairpins and off-targets. For low-abundance targets, test two to four candidate probes in separate reactions and keep the best performer.
2. Crosslinking scope. Psoralens primarily form interstrand crosslinks at opposing uridines (AU/UA). Interactions lacking appropriately positioned U's may be under-represented.
3. Positive/negative controls. AMT-treated (+AMT) and PBS-treated (−AMT) samples are UVA-irradiated and processed in parallel. Additional controls include a no-probe/beads-only control (bead/RNA stickiness), a scrambled-probe control (capture specificity), and a −UVA control (to confirm AMT is UVA-dependent) (*see* **Note 1**).

3.2 AMT Crosslinking of Cells

1. Grow cells (e.g., 786-O) on 10–15 cm plate at 37°C with 5% CO_2 to the desired density; harvest 2×10^7 cells per condition. Wash once with ice-cold PBS, pellet at 300–500 × *g* for 5 min at 4°C, discard the supernatant, and proceed to resuspension.
2. Prepare 0.5 mg/mL AMT in PBS by dissolving AMT at 1 mg/mL in nuclease-free water, then diluting 1:1 with 2× PBS. Place on ice in the dark.
3. Resuspend the cell pellet in 4 mL ice-cold 0.5 mg/mL AMT (+AMT) or PBS (−AMT). Incubate 15 min on ice, in the dark.
4. Transfer to a pre-chilled 10 cm plate on ice, cover with a 0.3 mm glass plate, and irradiate at 350 nm ~4 cm below the bulb for a total of 15 min UVA (e.g., 450 s twice), gently swirling once midway to ensure even exposure.
5. Transfer cells to pre-chilled tubes; pellet at 1000 × *g*, 5 min at 4°C. Discard supernatant. Pause point: Pellets can be snap-frozen at −80°C.

3.3 Lysis and RNA Purification

1. Lyse the pellet in 500 μL Cell Lysis Buffer supplemented with 200 U SUPERase·In RNase inhibitor. Incubate on ice for 10 min (*see* **Note 2**).
2. Add 20 U TURBO DNase, mix gently, and incubate 15 min at 37°C.
3. Add Proteinase K to 0.25 mg/mL and incubate 30 min at 37°C.

4. Purify RNA using the RNeasy Mini Kit (Qiagen) according to the manufacturer's instructions and elute in nuclease-free water (*see* **Note 3**).

3.4 Probe Hybridization

1. For each reaction, denature 2 μg input RNA and 50 pmol 5′-biotinylated ASO separately at 85°C for 3 min in 5 μL nuclease-free water; place immediately on ice and briefly spin (*see* **Note 4**).
2. Combine the RNA and probe in 300 μL LiCl Hybridization Buffer pre-equilibrated to 55°C and incubate at 55°C for 2–3 h at 1000 rpm in a thermomixer (1.5–2.0 mm, orbit). Use 3 h for low-abundance or highly structured targets (*see* **Note 5**).

3.5 Capture on Streptavidin Beads

1. Pre-wash 50 μL Dynabeads MyOne Streptavidin C1 three times with LiCl Hybridization Buffer; resuspend in 50 μL of the same buffer (*see* **Note 6**).
2. Add beads to the hybridization reaction and incubate at 37°C for 45 min with shaking, briefly resuspending every ~10 min to keep beads in suspension.
3. Place the tube on a magnetic rack and remove the supernatant completely; proceed to stringent washes.

3.6 Stringent Washes

1. Wash beads 3× with 250 μL Low Stringency Wash Buffer, each 5 min at 60°C with gentle agitation.
2. Wash beads 3× with 250 μL High Stringency Wash Buffer, each 5 min at 60°C with gentle agitation.
3. After the final wash, remove residual liquid completely (quick spin, return to magnet, wick away droplets) (*see* **Note 7**).

3.7 RNase H Elution and RNA Clean-Up

1. Rinse beads once with 250 μL RNase H Elution Buffer, place on the magnet, and remove the wash. Add 25–55 μL RNase H Elution mix containing RNase H at 2.5–5 units per reaction. Incubate 20–30 min at 37°C with shaking, then collect the eluate (*see* **Note 8**).
2. Purify RNA using TRIzol followed by RNA Clean & Concentrator-5 columns and elute in 16 μL nuclease-free water. Pause point: Store eluates at −80°C.

3.8 cDNA Synthesis and RT-qPCR

1. Synthesize cDNA with 5× PrimeScript RT Master Mix. For each sample (+AMT capture, −AMT capture, matched input), assemble a 10 μL RT reaction at 1×: 8 μL RAP-RNA eluate + 2 μL 5× Master Mix. Incubate 37°C for 15 min, then 85°C for 5 s; hold at 4°C (*see* **Note 9**). Dilute cDNA 1:3 in nuclease-free water before qPCR (*see* **Note 10**).
2. Quantify recovery by RT-qPCR using primers that flank, but do not overlap, the capture-probe binding site (outside the probe

footprint) to avoid probe carryover artifacts. Normalization (input-normalized, recommended): save an input aliquot (fraction f, e.g., f=0.01 for 1%). For each condition (+AMT and −AMT), compute %Input = 100 × f × 2^(Ctinput−Ctcapture) then report crosslink-dependent enrichment as Enrichment = (%Input_ + AMT)/(%Input_−AMT) (*see* **Notes 11 and 12**).

4 Notes

1. Maintain an RNase-free workflow by using RNase-free plastics/solutions and wiping benches and tools with RNase decontaminant. Handle AMT under low light with appropriate PPE, and shield samples and personnel from UVA during irradiation.
2. After lysis, if the lysate is viscous, shear it by passing 3–5× through a 26G needle while keeping on ice and avoiding bubbles. This reduces genomic DNA–driven viscosity and improves DNase/Proteinase K access.
3. An optional step (for poly(A)+ targets): If both the bait and intended RT-qPCR target are polyadenylated, perform poly(A)+ selection with oligo-dT25 magnetic beads to reduce rRNA background, then perform DNase treatment (if needed) and a final cleanup before proceeding.
4. If signal is low, increase RNA to 5–10 μg and scale the probe to ~25 pmol per μg RNA, increasing hybridization volume and bead amount proportionally.
5. An optional step (for probe-on-beads format): Pre-load the biotinylated ASO onto streptavidin beads in B&W buffer, wash beads into pre-warmed LiCl Hybridization Buffer, then add RNA and incubate 2–3 h at 55°C.
6. An optional step (for bead blocking): Pre-wash 50 μL Dynabeads MyOne Streptavidin C1 in B&W buffer, then block in 500 μL B&W supplemented with 1 mg/mL BSA (RNase-free) and 0.1 mg/mL yeast tRNA for 15–30 min at room temperature with rotation. Wash beads 2–3× with LiCl Hybridization Buffer to exchange buffer, resuspend, and proceed to capture.
7. Pre-warm wash buffers to 60°C; detergent/urea carryover can reduce RNase H efficiency.
8. An optional step (for post-elution DNase cleanup): Add 1 μL TURBO DNase to eluate; incubate 20 min at 37°C (1200 rpm in thermomixer) to remove any residual DNA, then proceed to RNA cleanup.

9. For very low-abundance targets, gene-specific priming may increase sensitivity. Include a −RT control (omit Master Mix; replace with water) and no-template control (NTC), and carry the −RT through qPCR to confirm no DNA/probe carryover.
10. An optional step (for post-RT cleanup): Treat cDNA with RNase H (37°C for 10 min) to remove RNA in RNA:DNA hybrids, then dilute cDNA 1:3 in nuclease-free water before qPCR.
11. Use the same primer pair for input and capture. Keep input and capture on the same cDNA dilution and qPCR input volume **(D = 1)**. If input is diluted extra by a factor D relative to capture (or different cDNA volume is added to qPCR), correct as **Ct_input, adj** = Ct_input − $\log_2$(D) and use **Ct_input, adj** in the % input formula. In this protocol, we reverse-transcribed 50% of the capture eluate (r_capture = 0.5); to report total % input captured, multiply the calculated % input by 1/r_capture (i.e., ×2).
12. Troubleshooting: If non-specific recovery is high (−AMT elevated), extend wash times to 8–10 min and/or raise the hybridization temperature to 60°C; also test an additional probe and include a scrambled-probe control. If the signal is low in both +AMT and −AMT, verify probe integrity/concentration and streptavidin bead activity (e.g., with a biotinylated oligo test), and confirm RNase H activity on a control RNA:DNA duplex.

Acknowledgments

We thank Nathan Smith for helpful comments on the manuscript. S.W., B.F., and J.-S.L. were supported by start-up funds from Rosalind Franklin University of Medicine and Science.

References

1. Guil S, Esteller M (2015) RNA-RNA interactions in gene regulation: the coding and noncoding players. Trends Biochem Sci 40(5):248–256. https://doi.org/10.1016/j.tibs.2015.03.001
2. Kornienko IV, Aramova OY, Tishchenko AA, Rudoy DV, Chikindas ML (2024) RNA stability: a review of the role of structural features and environmental conditions. Molecules 29(24):5978. https://doi.org/10.3390/molecules29245978
3. Ye R, Zhao H, Wang X, Xue Y (2024) Technological advancements in deciphering RNA-RNA interactions. Mol Cell 84(19):3722–3736. https://doi.org/10.1016/j.molcel.2024.06.036
4. Xue Y (2022) Architecture of RNA–RNA interactions. Curr Opin Genet Dev 72:138–144. https://doi.org/10.1016/j.gde.2021.11.007
5. Taft RJ, Pang KC, Mercer TR, Dinger M, Mattick JS (2010) Non-coding RNAs: regulators of disease. J Pathol 220(2):126–139. https://doi.org/10.1002/path.2638
6. Lee RC, Feinbaum RL, Ambros V (1993) The C. elegans heterochronic gene lin-4 encodes small RNAs with antisense complementarity

to lin-14. Cell 75(5):843–854. https://doi.org/10.1016/0092-8674(93)90529-Y
7. Kopp F, Mendell JT (2018) Functional classification and experimental dissection of long noncoding RNAs. Cell 172(3):393–407. https://doi.org/10.1016/j.cell.2018.01.011
8. Ramon YCS, Segura MF, Hummer S (2019) Interplay between ncRNAs and cellular communication: a proposal for understanding cell-specific signaling pathways. Front Genet 10:281. https://doi.org/10.3389/fgene.2019.00281
9. Wahl MC, Will CL, Luhrmann R (2009) The spliceosome: design principles of a dynamic RNP machine. Cell 136(4):701–718. https://doi.org/10.1016/j.cell.2009.02.009
10. Lui L, Lowe T (2013) Small nucleolar RNAs and RNA-guided post-transcriptional modification. Essays Biochem 54:53–77. https://doi.org/10.1042/bse0540053
11. Bartel DP (2004) MicroRNAs: genomics, biogenesis, mechanism, and function. Cell 116(2):281–297. https://doi.org/10.1016/S0092-8674(04)00045-5
12. Jonas S, Izaurralde E (2015) Towards a molecular understanding of microRNA-mediated gene silencing. Nat Rev Genet 16(7):421–433. https://doi.org/10.1038/nrg3965
13. Gabryelska MM, Conn SJ (2023) The RNA interactome in the Hallmarks of Cancer. Wiley Interdiscip Rev: RNA 14(5):e1786. https://doi.org/10.1002/wrna.1786
14. Li Y, Sun S (2025) RNA dysregulation in neurodegenerative diseases. EMBO J 44(3):613–638. https://doi.org/10.1038/s44318-024-00352-6
15. Duenas A, Exposito A, Aranega A, Franco D (2019) The role of non-coding RNA in congenital heart diseases. J Cardiovasc Dev Dis 6(2):15. https://doi.org/10.3390/jcdd6020015
16. Balas MM, Hartwick EW, Barrington C, Roberts JT, Wu SK, Bettcher R, Griffin AM, Kieft JS, Johnson AM (2021) Establishing RNA-RNA interactions remodels lncRNA structure and promotes PRC2 activity. Sci Adv 7(16):eabc9191. https://doi.org/10.1126/sciadv.abc9191
17. Clemson CM, Hutchinson JN, Sara SA, Ensminger AW, Fox AH, Chess A, Lawrence JB (2009) An architectural role for a nuclear noncoding RNA: NEAT1 RNA is essential for the structure of paraspeckles. Mol Cell 33(6):717–726. https://doi.org/10.1016/j.molcel.2009.01.026
18. Derrien T, Johnson R, Bussotti G, Tanzer A, Djebali S, Tilgner H, Guernec G, Martin D, Merkel A, Knowles DG, Lagarde J, Veeravalli L, Ruan X, Ruan Y, Lassmann T, Carninci P, Brown JB, Lipovich L, Gonzalez JM, Thomas M, Davis CA, Shiekhattar R, Gingeras TR, Hubbard TJ, Notredame C, Harrow J, Guigo R (2012) The GENCODE v7 catalog of human long noncoding RNAs: analysis of their gene structure, evolution, and expression. Genome Res 22(9):1775–1789. https://doi.org/10.1101/gr.132159.111
19. Cabili MN, Dunagin MC, McClanahan PD, Biaesch A, Padovan-Merhar O, Regev A, Rinn JL, Raj A (2015) Localization and abundance analysis of human lncRNAs at single-cell and single-molecule resolution. Genome Biol 16(1):20. https://doi.org/10.1186/s13059-015-0586-4
20. Tong C, Yin Y (2021) Localization of RNAs in the nucleus: cis- and trans- regulation. RNA Biol 18(12):2073–2086. https://doi.org/10.1080/15476286.2021.1894025
21. Wang J, Han X, Yuan Y, Gu H, Liao X, Jiang M (2022) The value of dysregulated LncRNAs on clinicopathology and survival in non-small-cell lung cancer: a systematic review and meta-analysis. Front Genet 13:821675. https://doi.org/10.3389/fgene.2022.821675
22. Bridges MC, Daulagala AC, Kourtidis A (2021) LNCcation: lncRNA localization and function. J Cell Biol 220(2):e202009045. https://doi.org/10.1083/jcb.202009045
23. Uszczynska-Ratajczak B, Lagarde J, Frankish A, Guigo R, Johnson R (2018) Towards a complete map of the human long non-coding RNA transcriptome. Nat Rev Genet 19(9):535–548. https://doi.org/10.1038/s41576-018-0017-y
24. Besançon W, Wagner R (1999) Characterization of transient RNA-RNA interactions important for the facilitated structure formation of bacterial ribosomal 16S RNA. Nucleic Acids Res 27(22):4353–4362. https://doi.org/10.1093/nar/27.22.4353
25. Lu Z, Chang HY (2018) The RNA base-pairing problem and base-pairing solutions. Cold Spring Harb Perspect Biol 10(12):a034926. https://doi.org/10.1101/cshperspect.a034926
26. Wang K, Liu F, Liu CY, An T, Zhang J, Zhou LY, Wang M, Dong YH, Li N, Gao JN, Zhao YF, Li PF (2016) The long noncoding RNA NRF regulates programmed necrosis and myocardial injury during ischemia and reperfusion by targeting miR-873. Cell Death Differ 23(8):1394–1405. https://doi.org/10.1038/cdd.2016.28
27. Bose M, Rankovic B, Mahamid J, Ephrussi A (2024) An architectural role of specific RNA-RNA interactions in oskar granules. Nat Cell Biol 26(11):1934–1942. https://doi.org/10.1038/s41556-024-01519-3
28. Schonberger B, Schaal C, Schafer R (2018) RNA interactomics: recent advances and remaining challenges. F1000Res 7:1824. https://doi.org/10.12688/f1000research.16146.1
29. Lu Z, Zhang Qiangfeng C, Lee B, Flynn Ryan A, Smith Martin A, Robinson James T, Davidovich

C, Gooding Anne R, Goodrich Karen J, Mattick John S, Mesirov Jill P, Cech Thomas R, Chang Howard Y (2016) RNA duplex map in living cells reveals higher-order transcriptome structure. Cell 165(5):1267–1279. https://doi.org/10.1016/j.cell.2016.04.028

30. Lu Z, Gong J, Zhang QC (2018) PARIS: Psoralen analysis of RNA interactions and structures with high throughput and resolution. Methods Mol Biol 1649:59–84. https://doi.org/10.1007/978-1-4939-7213-5_4
31. Sharma E, Sterne-Weiler T, O'Hanlon D, Blencowe Benjamin J (2016) Global mapping of human RNA-RNA interactions. Mol Cell 62(4):618–626. https://doi.org/10.1016/j.molcel.2016.04.030
32. Aw Jong Ghut A, Shen Y, Wilm A, Sun M, Lim Xin N, Boon K-L, Tapsin S, Chan Y-S, Tan C-P, Sim Adelene YL, Zhang T, Susanto Teodorus T, Fu Z, Nagarajan N, Wan Y (2016) In vivo mapping of Eukaryotic RNA interactomes reveals principles of higher-order organization and regulation. Mol Cell 62(4):603–617. https://doi.org/10.1016/j.molcel.2016.04.028
33. Ziv O, Gabryelska MM, Lun ATL, Gebert LFR, Sheu-Gruttadauria J, Meredith LW, Liu ZY, Kwok CK, Qin CF, MacRae IJ, Goodfellow I, Marioni JC, Kudla G, Miska EA (2018) COMRADES determines in vivo RNA structures and interactions. Nat Methods 15(10):785–788. https://doi.org/10.1038/s41592-018-0121-0
34. Engreitz JM, Sirokman K, McDonel P, Shishkin AA, Surka C, Russell P, Grossman SR, Chow AY, Guttman M, Lander ES (2014) RNA-RNA interactions enable specific targeting of noncoding RNAs to nascent Pre-mRNAs and chromatin sites. Cell 159(1):188–199. https://doi.org/10.1016/j.cell.2014.08.018
35. Maser RL, Calvet JP (1989) U3 small nuclear RNA can be psoralen-cross-linked in vivo to the 5' external transcribed spacer of pre-ribosomal-RNA. Proc Natl Acad Sci U S A 86(17):6523–6527. https://doi.org/10.1073/pnas.86.17.6523
36. Ko J, Lee Y, Park I, Cho B (2001) Identification of a structural motif of 23S rRNA interacting with 5S rRNA. FEBS Lett 508(3):300–304. https://doi.org/10.1016/S0014-5793(01)03068-X
37. Paillart JC, Skripkin E, Ehresmann B, Ehresmann C, Marquet R (1996) A loop-loop "kissing" complex is the essential part of the dimer linkage of genomic HIV-1 RNA. Proc Natl Acad Sci U S A 93(11):5572–5577. https://doi.org/10.1073/pnas.93.11.5572
38. Ul Haq I, Müller P, Brantl S (2024) In vitro methods for the investigation of sRNA-mRNA interactions in bacillus subtilis. In: Arluison V, Valverde C (eds) Bacterial regulatory RNA: methods and protocols. Springer US, New York, NY, pp 145–174. https://doi.org/10.1007/978-1-0716-3565-0_8
39. Ha T, Enderle T, Ogletree DF, Chemla DS, Selvin PR, Weiss S (1996) Probing the interaction between two single molecules: fluorescence resonance energy transfer between a single donor and a single acceptor. Proc Natl Acad Sci U S A 93(13):6264–6268. https://doi.org/10.1073/pnas.93.13.6264
40. Salim N, Lamichhane R, Zhao R, Banerjee T, Philip J, Rueda D, Feig AL (2012) Thermodynamic and kinetic analysis of an RNA kissing interaction and its resolution into an extended duplex. Biophys J 102(5):1097–1107. https://doi.org/10.1016/j.bpj.2011.12.052
41. Weeks KM, Mauger DM (2011) Exploring RNA structural codes with SHAPE chemistry. Acc Chem Res 44(12):1280–1291. https://doi.org/10.1021/ar200051h
42. Christy TW, Giannetti CA, Houlihan G, Smola MJ, Rice GM, Wang J, Dokholyan NV, Laederach A, Holliger P, Weeks KM (2021) Direct mapping of higher-order RNA interactions by SHAPE-JuMP. Biochemistry 60(25):1971–1982. https://doi.org/10.1021/acs.biochem.1c00270
43. Moazed D, Noller HF (1986) Transfer RNA shields specific nucleotides in 16S ribosomal RNA from attack by chemical probes. Cell 47(6):985–994. https://doi.org/10.1016/0092-8674(86)90813-5
44. Moazed D, Noller HF (1989) Interaction of tRNA with 23S rRNA in the ribosomal A, P, and E sites. Cell 57(4):585–597. https://doi.org/10.1016/0092-8674(89)90128-1
45. Lai D, Meyer IM (2015) A comprehensive comparison of general RNA–RNA interaction prediction methods. Nucleic Acids Res 44(7):e61. https://doi.org/10.1093/nar/gkv1477
46. Song P-S, Tapley KJ (1979) Photochemistry and photobiology of psoralens. Photochem Photobiol 29(6):1177–1197. https://doi.org/10.1111/j.1751-1097.1979.tb07838.x
47. McHugh CA, Guttman M (2018) RAP-MS: a method to identify proteins that interact directly with a specific RNA molecule in cells. In: Gaspar I (ed) RNA detection: methods and protocols. Springer, New York, NY, pp 473–488. https://doi.org/10.1007/978-1-4939-7213-5_31
48. Engreitz JM, Pandya-Jones A, McDonel P, Shishkin A, Sirokman K, Surka C, Kadri S, Xing J, Goren A, Lander ES, Plath K, Guttman M (2013) The Xist lncRNA exploits three-dimensional genome architecture to spread across the

X chromosome. Science 341(6147):1237973. https://doi.org/10.1126/science.1237973

49. Lee J-S, Dan T, Zhang H, Cheng Y, Rehfeld F, Brugarolas J, Mendell JT (2025) An ultraconserved snoRNA-like element in long noncoding RNA CRNDE promotes ribosome biogenesis and cell proliferation. Mol Cell 85(8): 1543–1560.e1510. https://doi.org/10.1016/j.molcel.2025.03.006

50. Bhat P, Chow A, Emert B, Ettlin O, Quinodoz SA, Strehle M, Takei Y, Burr A, Goronzy IN, Chen AW, Huang W, Ferrer JLM, Soehalim E, Goh ST, Chari T, Sullivan DK, Blanco MR, Guttman M (2024) Genome organization around nuclear speckles drives mRNA splicing efficiency. Nature 629(8014): 1165–1173. https://doi.org/10.1038/s41586-024-07429-6

Chapter 15

Assay for In-Vitro Arginine Methylation of RNA Binding Proteins Followed by RNA Binding

Sakshi Dewasthale, Gayatri Mohanan and Purusharth I. Rajyaguru

Abstract

Arginine methylation is a prevalent post-translational modification that plays a critical role in regulating the function of RNA-binding proteins [RBPs]. RBPs are the largest class of arginine methylated proteins. Catalyzed by protein arginine methyltransferases [PRMTs], this modification influences RNA recognition, binding affinity, and the formation of RNA–protein complexes. This impacts diverse cellular processes such as RNA splicing, transport, stability, and translation. Arginine methylation of RGG motifs modulates RBP behavior, with implications for normal physiology and disease states, including neurodegeneration, cancer, and immune disorders. In vitro methylation assays combined with RNA binding studies offer a powerful approach to elucidate how arginine methylation affects RNA–protein interactions. These techniques allow precise manipulation of methylation states and controlled assessment of binding dynamics, providing mechanistic insights into regulatory pathways. The protocols described in this chapter outline methods for recombinant protein expression, enzymatic methylation, and RNA binding assays, offering adaptable tools for researchers studying the functional consequences of arginine methylation in RNA biology. Together, these methodologies provide a framework to explore the interplay between post-translational modifications and RNA regulation, deepening our understanding of cellular processes.

Keywords Arginine methylation, RGG-motif, RNA-binding, RNA-protein interaction

1 Introduction

Post-translational modifications [PTMs] of proteins are pivotal regulators of cellular physiology, influencing protein conformation, interactions, localization, stability, and enzymatic activity [1–4]. Among these, arginine methylation has gained significant attention as a conserved and dynamic modification that modulates the functions of various proteins across eukaryotic species [5]. Arginine methylation is catalyzed by the protein arginine methyltransferase [PRMT] family, which transfers methyl groups from the methyl donor S-adenosylmethionine [SAM] to the guanidino nitrogen atoms of arginine residues. This process results in the formation of monomethyl arginine [MMA], asymmetric dimethylarginine [ADMA], or

Je-Hyun Yoon (ed.), *RNA Labeling*, Methods in Molecular Biology, vol. 3047,
https://doi.org/10.1007/978-1-0716-5352-4_15,

symmetric dimethylarginine [SDMA], depending on the enzyme involved [6]. Nine PRMTs [PRMT1–9] have been identified in mammals. PRMTs have been classified into four categories depending upon their catalytic activity. "Type I" PRMTs catalyze the formation of MMA and ADMA; "type II" catalyzes MMA and SDMA formation; "type III" catalyzes MMA formation and "type IV" catalyze the formation of δ-MMA [7].

PRMT1, the predominant enzyme responsible for MMA and ADMA formation, plays crucial roles in transcriptional regulation, signal transduction, and RNA metabolism [8]. PRMT5, a type II enzyme that catalyzes SDMA, is essential in chromatin remodeling and splicing [9]. Other family members, such as PRMT7, PRMT6, and PRMT8, have specialized roles in development, stress response, and neural functions [10, 11]. The diversity of PRMTs and their substrates underscores the broad regulatory capacity of arginine methylation and its intricate involvement in cellular homeostasis.

RNA-binding proteins [RBPs] represent a vast and functionally diverse class of molecules that orchestrate post-transcriptional gene regulation by interacting with RNA molecules [12]. These proteins are involved in a wide array of processes, including pre-mRNA splicing, RNA transport, translation, degradation, and storage within membrane-less organelles such as stress granules and processing bodies [13, 14]. RBPs utilize defined structural motifs for RNA recognition, such as RNA recognition motifs [RRMs], K homology [KH] domains, zinc finger domains, and arginine–glycine–glycine [RGG/RG] boxes [15]. The RGG motif, in particular, is rich in arginine residues and is a preferential target for methylation by PRMTs. Methylation at these sites can affect RNA-binding affinity, phase separation behavior, and interactions with other proteins, thereby regulating the formation of RNA-protein complexes and influencing cellular stress responses [16].

The regulation of RNA binding by post-translational modifications such as arginine methylation is an important mechanism that allows cells to adapt to changing physiological and stress condition. Arginine methylation fine-tunes the behavior of RBPs by influencing their RNA affinity, subcellular localization, and ability to undergo phase separation [17]. For example, the RBP FUS [Fused in Sarcoma] is methylated by PRMT1 and PRMT8 at arginine residues within its RGG-rich domain [18]. This modification weakens FUS–RNA interactions and reduces its tendency to form condensates under normal conditions [19]. When FUS is hypomethylated, it accumulates in the cytoplasm and forms aggregates, a pathological feature commonly observed in amyotrophic lateral sclerosis [ALS] and frontotemporal dementia [FTD] [20]. Similarly, hnRNPA1, another RGG-containing protein, is methylated by PRMT1, which facilitates its nuclear import through Transportin-1 and maintains its solubility [21]. Loss of methylation leads to its cytoplasmic mislocalization and aggregation, contributing to ALS [22].

Methylation also plays a key role in regulating other RBPs. EWSR1 is methylated by PRMT1, which influences its RNA-binding activity and coupling between transcription and splicing [23]. Altered methylation of EWSR1 has been linked to transcriptional dysregulation in Ewing sarcoma and to FTD-like phenotypes [24]. Symmetric di-methylation of Sm proteins by PRMT5 is required for the proper assembly of spliceosomal snRNPs; defects in this modification can disrupt pre-mRNA splicing and have been associated with spinal muscular atrophy [SMA] [25]. In addition, methylation of TDP-43 by PRMT1 and PRMT8 affects its phase separation and nucleocytoplasmic distribution [26]. Beyond neurodegeneration, methylation of RBPs such as hnRNPK and Sam68 has been shown to modulate viral RNA processing and replication. This influencing host–virus interactions in infections like HIV-1 and hepatitis B [27, 28]. Together, these examples highlight the multifaceted role of arginine methylation in controlling RBP behavior—acting as a molecular switch that regulates RNA binding, localization, and condensate formation.

RNA binding itself is governed by the interplay between sequence-specific recognition and structural compatibility. RBPs bind target RNAs through electrostatic interactions, hydrogen bonds, and base-stacking interactions, which can be fine-tuned by PTMs [29]. Arginine methylation modulates the charge distribution and steric properties of arginine residues. This can directly affect the binding between RBPs and their RNA targets. This regulation is critical for gene expression [30].

A wide range of techniques is employed to study RNA–protein interactions in vitro, including electrophoretic mobility shift assays [EMSAs], fluorescence anisotropy, surface plasmon resonance [SPR], and crosslinking immunoprecipitation [CLIP] methodologies [31–34]. These techniques enable the quantification of binding kinetics, the identification of binding sites, and the assessment of affinity changes upon modification or mutation [35]. Importantly, in vitro RNA-binding assays conducted in combination with enzymatic methylation reactions allow researchers to precisely define the impact of arginine methylation on RBP function, independent of the cellular processes [34].

In vitro arginine methylation assays provide a powerful platform to recapitulate and dissect the enzymatic activity of PRMTs, enabling the modification of recombinant RBPs under defined conditions. These assays allow manipulation of substrate concentration, methyl donor availability, and enzyme variants, thus facilitating detailed structure–function analyses. When paired with RNA binding assays, this approach provides direct evidence for the influence of arginine methylation on RNA recognition and binding dynamics.

Given the increasing interest in arginine methylation as a regulatory mechanism in RNA biology, robust methodologies are essential for the investigation of methylation-dependent interactions. In this chapter, we present a comprehensive guide for performing in vitro arginine methylation of RNA-binding proteins followed by RNA binding assays. The protocols include steps for recombinant protein expression and purification, enzymatic methylation reactions using purified PRMTs, and binding assessment through electrophoretic mobility shift assays.

Through these approaches, researchers can gain mechanistic insights into how arginine methylation modulates RNA binding, elucidate the roles of PRMTs in cellular function, and explore therapeutic avenues for diseases linked to RBP dysregulation. The protocols and considerations provided in this chapter aim to equip scientists with the tools necessary to explore the molecular crosstalk between post-translational modifications and RNA biology, offering new perspectives on gene regulation and cellular adaptation.

2 Materials

2.1 Protein Purification Using Ni-NTA Chromatography

1. Ni-NTA agarose beads [G Biosciences, Cat# 786-940].
2. Lysozyme [10 μg/mL], [Sigma, Cat# L6876].

   Buffer Compositions:
3. Lysis buffer [50 mM NaH_2PO_4, pH 8.0, 300 mM NaCl, 10 mM imidazole, pH 8.0, 1× Protease inhibitor cocktail,1 μ g/mL RNaseA].
4. Wash buffer [50 mM NaH_2PO_4, pH 8.0, 300 mM NaCl with 20 mM imidazole, pH 8.0 for wash 1, 35 mM imidazole, pH 8.0 for wash 2, and 50 mM imidazole, pH 8.0 for wash 3].
5. Elution buffer [50 mM NaH_2PO_4, pH 8.0, 300 mM NaCl, 500 mM imidazole, pH 8.0].
6. Dialysis buffer [10 mM Tris–Cl pH 7.5, 100 mM NaCl, 20% glycerol, 1 mM DTT].

2.2 In-Vitro Methylation

1. Recombinant proteins-Sbp1/Scd6 and Hmt1.
2. S-adenosylmethionine [SAM, 2 mM] [NEB, Cat# B9003S].

   Buffer Composition:
3. Methylation buffer [100 mM Tris pH 8.0, 200 mM NaCl, 2 mM EDTA, 1 mM DTT].

2.3 RNA-Binding by Electrophoretic Mobility Shift Assay [EMSA]

1. Purified His-Scd6-FLAG protein [unmethylated or methylated].
2. In vitro transcribed RNA fragments.
3. 2.5% agarose gel in 1× MOPS buffer.
4. Ethidium bromide for staining.

Buffer Composition:

5. Binding buffer: [10 mM Tris pH 8.0, 50 mM NaCl, 0.05% NP-40, 6% glycerol, 1 mM DTT, 2 µg/mL BSA, 0.5 µg/mL tRNA].

3 Methods

Arginine methylation is an important post-translational modification that fine-tunes the RNA-binding activity and function of many RNA-binding proteins (RBPs). The methods described here allow a direct comparison of how methylation affects the RNA-binding behavior of proteins such as Scd6 and Sbp1. By combining in vitro methylation with electrophoretic mobility shift assays, this approach provides a straightforward way to dissect how Hmt1-mediated methylation influences RBP–RNA interactions.

3.1 Protein Purification Using Ni-NTA Chromatography

3.1.1 Growing Cells and Harvesting

1. Inoculate 500 mL of LB broth with 0.1 OD [600 nm] from overnight culture and incubate in the shaker incubator at 37°C.
2. At 0.5 OD, induce cells with 1 mM IPTG for 4 h at 37°C.
3. Harvest cells at 8000 × *g* for 5 min.
4. Store pellets in −80°C or use immediately for protein purification.

3.1.2 Cell Lysis

1. Resuspend the cell pellet in 10 mL of lysis buffer.
2. Add lysozyme to a final concentration of 10 µg/mL and DTT to 1 mM.
3. Incubate the mixture for 30 min on ice to allow efficient cell lysis.
4. Sonicate for nine cycles [10 s on and 10 s off, 35% amplitude]
5. Centrifuge at 20,000 × *g* for 15 min.
6. Take the clear supernatant in a 15 mL centrifuge tube.

3.1.3 Binding to Ni-NTA Agarose

1. Add 500 µL Ni-NTA agarose beads 1:1 slurry, equilibrated in the lysis buffer to the lysate. [No imidazole for Ni-NTA beads equilibration].
2. Incubate the mixture on a nutator for 2 h at 4°C to facilitate binding between the His-tagged protein and the beads.

3.1.4 Washing

1. Spin the tubes at 200 × *g* at 4°C for 1 min.
2. Remove the supernatant. [Can store the lysate with unbound protein to repeat the binding if the yield is low.]

3. Wash the beads with 10 mL of wash buffer for 10 min on a nutator at 4°C to remove nonspecifically bound proteins and contaminants.
4. Spin the tubes at 200 × *g* at 4°C for 1 min.
5. Repeat washes 2 and 3 similarly.
6. More washes can be given depending on the protein quality upon elution.

3.1.5 Elution

1. Elute the bound protein by incubating the beads with elution buffer containing 500 mM imidazole.
2. Collect the eluted protein carefully and keep on ice.

3.1.6 Dialysis

1. Dialyze the eluted protein against dialysis buffer at 4°C.
2. Perform dialysis overnight

3.1.7 Storage

1. Store the purified protein at −80°C in 50 μL aliquots.

Figure 1 shows a CBB stained SDS-PAGE gel with purified Hmt1, Sbp1, and Scd6. Notes 1 provide important precautions for the successful execution of this protocol.

3.2 In-Vitro Methylation

3.2.1 Preparation of Reaction Mixture

1. In a 1.5 mL centrifuge tube, add 5 μg of recombinant His-Scd6-FLAG or His-Sbp1-FLAG protein.
2. Add 5 μg of recombinant His-Hmt1 protein.
3. Add 10× methylation buffer to a final concentration of 1× and make up the reaction volume to 250 μL.

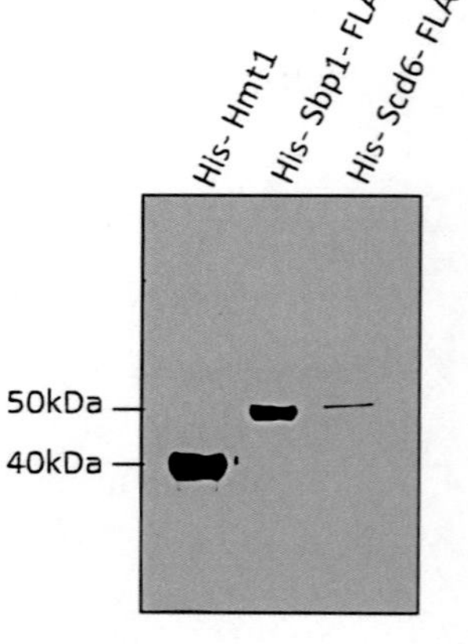

Fig. 1 Protein purification using Ni-NTA chromatography. Protein purification for His-Hmt1, His-Sbp1-FLAG, and His-Scd6-FLAG. Gel stained with CBB

4. Prepare two sets of reactions: one with 2 mM SAM and one without SAM, depending on the control and test conditions.

3.2.2 Initiating the Methylation Reaction

1. Mix the contents gently but thoroughly.
2. Incubate the reaction at 30°C for 2 h to allow methylation to occur.

   Post-Reaction Handling

3. After incubation, proceed with downstream analysis such as detection of arginine methylation by western blotting using arginine methylation-specific antibodies.

Figure 2 shows in-vitro methylation of Sbp1 and Scd6 by purified recombinant Hmt1. Notes 2 provide important precautions for the successful execution of this protocol.

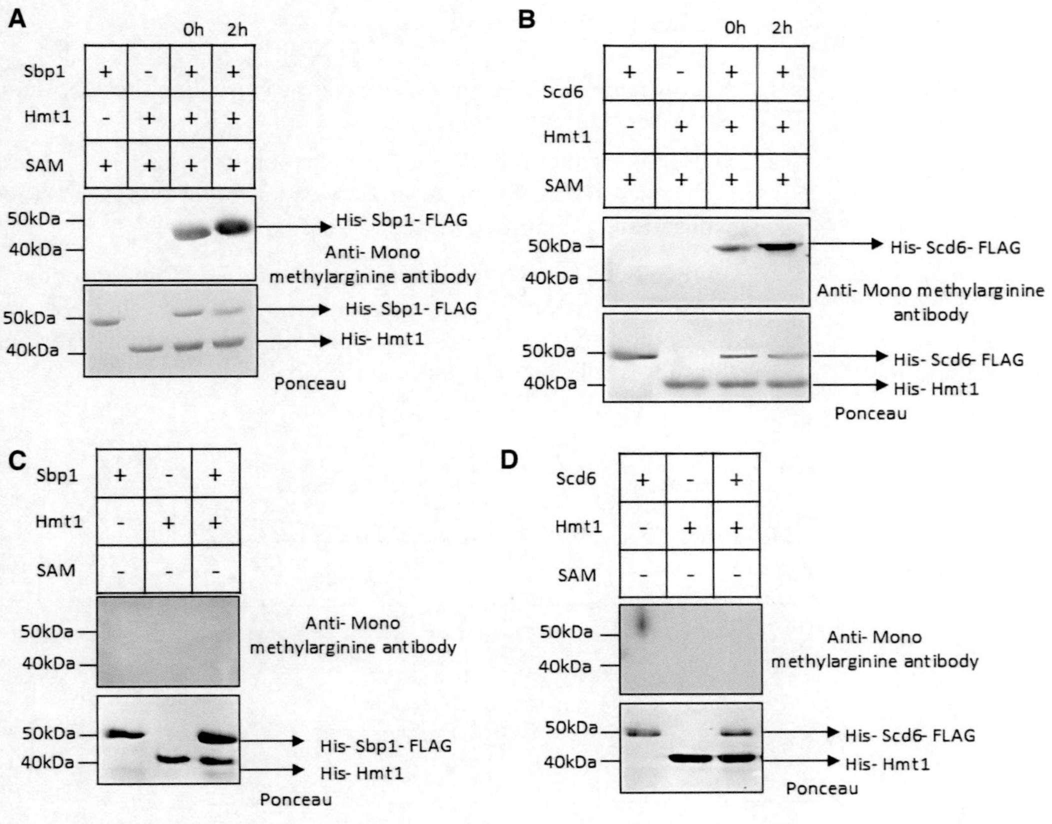

Fig. 2 In-vitro methylation assay. (**A**) In-vitro methylation for Sbp1, (**B**) in-vitro methylation for Scd6, (**C**) controls showing no methylation of Sbp1 in the absence of SAM after 2 h of incubation, (**D**) controls showing no methylation of Scd6 in the absence of SAM after 2 h of incubation

3.3 RNA-Binding by Electrophoretic Mobility Shift Assay [EMSA]

3.3.1 RNA Binding activity of Methylated and Unmethylated Scd6

1. Prepare separate series of reactions containing increasing concentrations of His-Scd6-FLAG protein [0.7, 1.4, 2.1, 2.8, 3.4, 4.1 μM], either in methylated or unmethylated form.
2. Add 0.94 μM of in vitro transcribed RNA to each reaction.
3. Set up the reaction in 1× Binding buffer.
4. Adjust the volume with dialysis buffer to 25 μL.

3.3.2 Binding Reaction

1. Incubate all reaction mixtures at 30°C for 30 min to allow binding between the protein and RNA.

3.3.3 Electrophoresis

1. Prepare a 2.5% agarose gel using 1× MOPS buffer.
2. Load each binding reaction onto the gel carefully to avoid disturbing the samples.
3. Run the gel under standard electrophoresis conditions suitable for RNA-protein complexes.

3.3.4 Visualization

1. Stain the gel with ethidium bromide to visualize the RNA and RNA-protein complexes.
2. Observe and compare the mobility shift patterns to assess differences in RNA binding between the various conditions [methylated vs unmethylated Scd6].

Figure 3 shows RNA binding (200-mer 5′UTR fragment of SRS2 mRNA) with increasing concentration of unmethylated and methylated protein (Scd6). Notes 3 provide important precautions for the successful execution of this protocol.

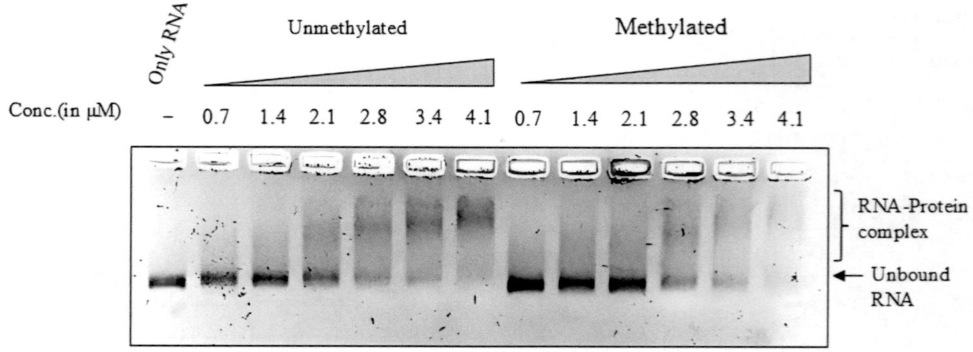

Fig. 3 RNA binding by electrophoretic mobility shift assay (EMSA). Showing increased RNA (200-mer 5′UTR fragment of *SRS2* mRNA) binding with increasing concentration of unmethylated protein (Scd6), and decreased RNA binding with increasing concentration of methylated protein (Scd6)

4 Notes

1. **Protein Purification**
   - Always work on ice during cell lysis, binding, washing, and dialysis steps to minimize proteolytic degradation and preserve protein activity.
   - Add protease inhibitors freshly before cell lysis and keep them on ice throughout the procedure.
   - Use clean, RNase-free equipment when handling samples for EMSA or RNA binding assays to prevent contamination and degradation of RNA.
   - Ensure that imidazole concentration during elution is optimized; excessively high concentrations can lead to protein precipitation or carryover contaminants.
2. **In-Vitro Arginine Methylation**
   - Use fresh SAM solutions, as SAM is sensitive to oxidation and degradation.
   - Avoid repeated freeze-thaw cycles of purified proteins, as this can lead to loss of activity or aggregation.
   - Confirm the integrity and concentration of proteins before starting the methylation reaction to ensure reproducibility.
   - Always include appropriate controls [with and without SAM] to validate that the observed effects are specific to methylation.
   - Perform reactions in low-binding tubes to minimize protein adsorption to the tube walls.
   - Note: Methylation is seen even at 0 h time point.
3. **Electrophoretic Mobility Shift Assay [EMSA]**
   - Ethidium bromide is a mutagen; handle with care, wear gloves, and dispose of staining solutions as hazardous waste according to institutional guidelines.
   - Run pilot gels to optimize the agarose concentration and electrophoresis time for the specific protein–RNA complexes being studied.
   - Avoid overheating the gel during electrophoresis, as excessive temperature can cause diffusion or degradation of RNA.
   - Include appropriate negative controls [e.g., reactions without protein or with unrelated RNA] to confirm the specificity of RNA binding.

References

1. Bah A, Forman-Kay JD (2016) Modulation of intrinsically disordered protein function by post-translational modifications. J Biol Chem 291(13):6696–6705
2. Duan G, Walther D (2015) The roles of post-translational modifications in the context of protein interaction networks. PLOS Comput Biol 11(2):e1004049
3. Zhong Q, Xiao X, Qiu Y, Xu Z, Chen C, Chong B et al. (2023) Protein posttranslational modifications in health and diseases: functions, regulatory mechanisms, and therapeutic implications. MedComm [2020] 4(3):e261
4. Lee JM, Hammarén HM, Savitski MM, Baek SH (2023) Control of protein stability by post-translational modifications. Nat Commun 14(1):201
5. Bedford MT, Richard S (2005) Arginine methylation an emerging regulator of protein function. Mol Cell 18(3):263–272
6. Yang Y, Bedford MT (2013) Protein arginine methyltransferases and cancer. Nat Rev Cancer 13(1):37–50
7. Tewary SK, Zheng YG, Ho MC (2019) Protein arginine methyltransferases: insights into the enzyme structure and mechanism at the atomic level. Cell Mol Life Sci 76(15):2917–2932
8. Wolf SS (2009) The protein arginine methyltransferase family: an update about function, new perspectives and the physiological role in humans. Cell Mol Life Sci 66(13):2109–2121
9. Stopa N, Krebs JE, Shechter D (2015) The PRMT5 arginine methyltransferase: many roles in development, cancer and beyond. Cell Mol Life Sci 72(11):2041–2059
10. Xie B, Yu J, Chen C, Shen T (2025) Protein arginine methyltransferases from regulatory function to clinical implication in central nervous system. Cell Mol Neurobiol 45:41
11. Halabelian L, Barsyte-Lovejoy D (2021) Structure and function of protein arginine methyltransferase PRMT7. Life 11(8):768
12. Glisovic T, Bachorik JL, Yong J, Dreyfuss G (2008) RNA-binding proteins and post-transcriptional gene regulation. FEBS Lett 582(14):1977–1986
13. Fredericks AM, Cygan KJ, Brown BA, Fairbrother WG (2015) RNA-binding proteins: splicing factors and disease. Biomolecules 5(2):893–909
14. Yan Y, Gan J, Tao Y, Okita TW, Tian L (2022) RNA-binding proteins: the key modulator in stress granule formation and abiotic stress response. Front Plant Sci 13:882596. https://doi.org/10.3389/fpls.2022.882596
15. Fang Y, Liu X, Liu Y, Xu N (2024) Insights into the mode and mechanism of interactions between RNA and RNA-binding proteins. Int J Mol Sci 25(21):11337
16. Chowdhury MN, Jin H (2023) The RGG motif proteins: interactions, functions, and regulations. Wiley Interdiscip Rev: RNA 14(1):e1748
17. Guccione E, Richard S (2019) The regulation, functions and clinical relevance of arginine methylation. Nat Rev Mol Cell Biol 20(10):642–657
18. Scaramuzzino C, Monaghan J, Milioto C, Lanson NA, Maltare A, Aggarwal T et al. (2013) Protein arginine methyltransferase 1 and 8 interact with FUS to modify its sub-cellular distribution and toxicity in vitro and in vivo. PLOS One 8(4):e61576
19. Hofweber M, Hutten S, Bourgeois B, Spreitzer E, Niedner-Boblenz A, Schifferer M et al. (2018) Phase separation of FUS is suppressed by its nuclear import receptor and arginine methylation. Cell 173(3):706–719.e13
20. Kim SC, Mitchell SJ, Qamar S, Whitcomb DJ, Ruepp MD, St George-Hyslop P et al. (2023) Mimicking hypomethylation of FUS requires liquid–liquid phase separation to induce synaptic dysfunctions. Acta Neuropathol Commun 11:199
21. Shen EC, Henry MF, Weiss VH, Valentini SR, Silver PA, Lee MS (1998) Arginine methylation facilitates the nuclear export of hnRNP proteins. Genes Dev 12(5):679–691
22. Clarke JP, Thibault PA, Salapa HE, Levin MC (2021) A comprehensive analysis of the role of hnRNP A1 function and dysfunction in the pathogenesis of neurodegenerative disease. Front Mol Biosci 8:659610
23. Paronetto MP (2013) Ewing sarcoma protein: a key player in human cancer. Int J Cell Biol 2013:642853
24. Li M, Chen CW (2022) Epigenetic and transcriptional signaling in ewing sarcoma—disease etiology and therapeutic opportunities. Biomedicines 10(6):1325
25. Meister G, Eggert C, Bühler D, Brahms H, Kambach C, Fischer U (2001) Methylation of Sm proteins by a complex containing PRMT5 and the putative U snRNP assembly factor pICln. Curr Biol 11(24):1990–1994
26. Áikio M, Odeh HM, Wobst HJ, Lee BL, Chan Ú, Mauna JC et al. (2025) Opposing roles of p38α-mediated phosphorylation and PRMT1-mediated arginine methylation in driving TDP-43 proteinopathy. Cell Rep 44(1):115205

27. Bomsztyk K, Denisenko O, Ostrowski J (2004) hnRNP K: one protein multiple processes. BioEssays 26(6):629–638
28. Sánchez-Jiménez F, Sánchez-Margalet V (2013) Role of Sam68 in post-transcriptional gene regulation. Int J Mol Sci 14(12):23402–23419
29. Corley M, Burns MC, Yeo GW (2020) How RNA binding proteins interact with RNA: molecules and mechanisms. Mol Cell 78(1):9–29
30. Wang X, Zhu B, Winn R, Lu S, Wang H (2025) Functional dynamics of arginine mono- and di-methylation. Cells 14(11):796
31. Rio DC (2014) Electrophoretic mobility shift assays for RNA-protein complexes. Cold Spring Harb Protoc 2014(4):435–440
32. Ule J, Hwang HW, Darnell RB (2018) The future of cross-linking and immunoprecipitation [CLIP]. Cold Spring Harb Perspect Biol 10(8):a032243
33. Vo T, Paul A, Kumar A, Boykin DW, Wilson WD (2019) Biosensor-surface plasmon resonance: a strategy to help establish a new generation RNA-specific small molecules. Methods 167:15–27
34. Licatalosi DD, Ye X, Jankowsky E (2020) Approaches for measuring the dynamics of RNA-protein interactions. Wiley Interdiscip Rev: RNA 11(1):e1565
35. Katsamba PS, Park S, Laird-Offringa IA (2002) Kinetic studies of RNA-protein interactions using surface plasmon resonance. Methods 26(2):95–104

Part IV

Beyond RNA Labeling

Check for updates

Chapter 16

Fluorescence-Based mRNA Translation Reporter

Jemima Riess, Gunhild Sommer and Tilman Heise

Abstract

Herein, we describe a fluorescence reporter assay for investigating regulatory elements in the 5′-UTR of mRNA that modulate translational initiation. The canonical translation initiation process generally begins with the assembly of the eIF4F complex at the cap structure, the formation of the ternary complex, which is loaded with the initiator tRNA, and finally, the preinitiation complex. The preinitiation complex then scans the 5′-UTR to identify the authentic translation start site, which is typically characterized by an AUG codon within a sequence context known as the Kozak sequence. It is well established that the 5′-UTR varies extensively in length and contains regulatory, structural, and sequence elements. To investigate these elements, we use a green fluorescence protein (GFP)- and a luciferase-based mRNA reporter. These reporters can be quickly generated by ordering gBlocks or by PCR-based template generation for in vitro transcription, capping and polyadenylation, to produce synthetic mRNAs. These mRNAs can be transfected into cells, and expression can be monitored via GFP fluorescence or by measuring luciferase activity. In this chapter, we provide a step-by-step protocol for designing gBlocks, generating the reporter mRNA, transfecting the cells, and analyzing the results.

Keywords RNA-binding, Translation, 5′-untranslated region, Fluorescence, Reporter, RNA structure, RNA-binding protein, Cell-based assay, Luciferase

1 Introduction

Messenger RNA (mRNA) provides the genetic information necessary for protein synthesis and is comprised of a 5′-untranslated region (5′-UTR), an open reading frame (ORF), and a 3′-untranslated region (3′-UTR). During the canonical translation initiation process, the eIF4 complex forms at the cap structure at the 5′ end of the mRNA. It then recruits the ternary complex and the small ribosomal subunits to form the preinitiation complex (PIC). The PIC scans the 5′-UTR until it recognizes an AUG start codon within a favorable sequence context referred to as the Kozak consensus sequence [1–4]. In contrast, noncanonical cap-independent translation is important under stress conditions or viral infection when canonical cap-dependent translation is impaired [5, 6]. Under cellular stress conditions, kinases can

Je-Hyun Yoon (ed.), *RNA Labeling*, Methods in Molecular Biology, vol. 3047,
https://doi.org/10.1007/978-1-0716-5352-4_16,

inactivate eIF2alpha, which leads to a shortage of the ternary complex and impairs cap-dependent translation [7, 8]. Similarly, the phosphorylation of 4E-binding proteins (4E-BPs) can inhibit cap-dependent translation [9–12]. Under such conditions, noncanonical translation initiation mechanisms can enable the translation of specific mRNAs. The 5′-UTR of mRNA varies greatly in length, structural features, and sequence motifs [1, 13, 14]. For example, long and structured 5′-UTRs may contain large structural elements that allow cap-independent recruitment of the translation machinery to viral and few cellular mRNAs containing internal ribosome entry sites (IRES; [15–20]). Long, structured 5′-UTRs can also mediate RNA ribosome shunting [21]. In this process, the scanning PIC encounters a structural element in the 5′-UTR and bypasses it ("shunt") without unfolding. Then, it proceeds with the scanning process until a suitable start site is encountered. Noncanonical translation can also initiate at structural elements recognized by eIF3 [22]. Other elements, such as upstream open reading frames (uORFs), allow translation of specific mRNAs when concentrations of the ternary complex are low [23]. Conversely, terminal oligopyrimidine (TOP) mRNAs are characterized by a short sequence motif located at the 5′-end that begins with a cytosine [24–26]. These mRNA sequences are characterized by polypyrimidine tracts at the 5′-end and encode a large number of ribosomal, translation initiation, and elongation factors. A major regulator of these TOP mRNAs is the mTOR signaling complex, which is sensitive to the nutritional content of cells [27–29]. Furthermore, small local structures can hinder translation. Although the scanning PIC contains RNA helicases (e.g., eIF4A) to unwind structural elements, additional proteins, such as RNA helicases and RNA chaperones, have been considered capable of promoting translation [30–36].

Moreover, several small sequence motifs have been described to regulate translation under various situations, often associated with tumor-promoting conditions. These include the cytosine-enriched regulator of translation (CERT; [37]), the pyrimidine-rich translational element (PRTE; [38, 39], and the translation inhibitory element (TIE; [40]).

This brief overview underscores the complexity of 5′-UTRs during translational initiation and the need for assays to study novel elements, characterize known elements, and determine the role of specific proteins in this process. These regulatory elements are commonly studied by transfecting reporter plasmids into cells. However, DNA transfections can trigger DNA-sensing pathways or cryptic promoters, or splice sites can be used. Strong promoters are often used, which can lead to high, nonphysiological expression levels. Once the plasmid has been transcribed in the nucleus, the mRNA undergoes extensive processing and must be exported to the cytoplasm. During this process, the element of interest may interfere with processing steps and intronless transcripts may not be efficiently exported to the cytoplasm. Therefore, transfecting mRNA provides a direct approach, bypassing the numerous cellular processes that new transcripts undergo before being incorporated into the translational machinery.

To study these processes, we present an easy-to-use assay protocol that can be employed to directly determine the functional significance of known and novel translational regulatory elements, as well as the role of specific factors in these processes. The four-step procedure includes (see Fig. 2a): (i) designing the reporter; (ii) ordering gBlocks or PCR-based generation of the template DNA; (iii) generating synthetic mRNAs; (iv) transfecting the cells of interest, and (v) analyzing expression using immunofluorescence microscopy or measuring luciferase expression.

2 Materials

Please bear in mind that you will be working with RNA molecules and that you should familiarize yourself with the necessary precautions to avoid RNA degradation (*see* **Note 1**). Prepare all solutions using nuclease-free molecular biology-grade water and ensure that purchased solutions are also nuclease-free.

1. Nuclease free molecular biology-grade water.
2. gBlocks (see Fig. 1):
   a. T7 Pro control reporter nanoLuc destabilized (769 nts, con_nanoLuc).
   b. T7 Pro test reporter nanoLuc destabilized (763 nts, test_nanoLuc).
   c. T7 Pro control destabilized GFP (970 nts, con_destaGFP).
3. DNA oligonucleotides (see Table 1, see Fig. 1)
   a. naLuc_repo-FP.
   b. naLuc_repo-RP.
   c. T7fireLucPCR FP (see Note 2).
   d. T7fireLucPCR RP.
4. Thermocycler.
5. PCR tubes or plates.
6. pmirGLO Dual-Luciferase miRNA Target Expression Vector.
7. Q5 high Fidelity Polymerase.
8. Qiaquick® PCR Purification Kit.
9. Nanodrop 2000.
10. Agarose.
11. MOPS.
12. Gel-Red.
13. 37% Formaldehyde.

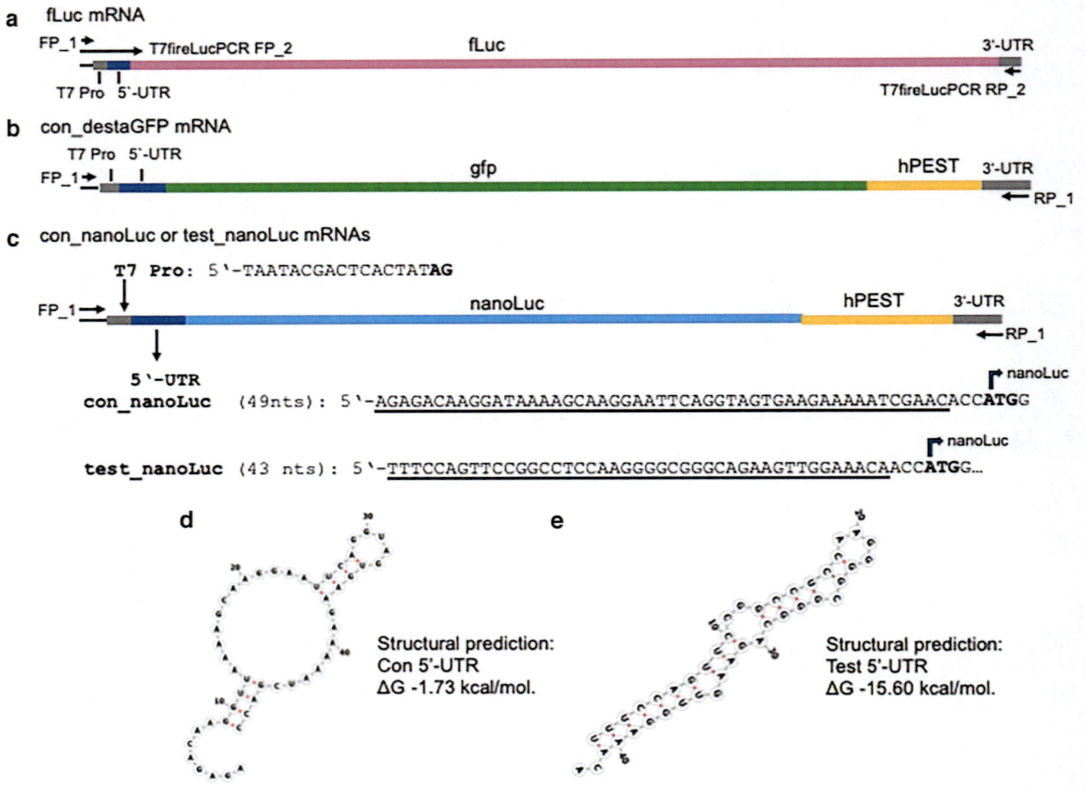

Fig. 1 The effect of the structure of the 5′UTR on translation was evaluated using different mRNA reporters. (**a**) Firefly Luciferase (fLuc) mRNA used for normalization. Primers T7firelucPCR_FP_2 (*see* **Notes 2 and 3**) and T7firelucPCR_RP_2 were used to generate the DNA template for this mRNA construct. The fLuc mRNA contains the same 5′-UTR as con_nanoLuc and encodes Firefly luciferase. The protein is not destabilized by an hPEST sequence. (**b**) The con_destaGFP mRNA was used to optimize the following transfection parameters: cell number, RNA amount, and Lipofectamine amount. The template was amplified from a gBlock using primers FP_1 and RP_1. The Con_destaGFP mRNA encodes the green fluorescence protein (GFP) fused to an hPEST sequence, which promotes rapid protein turnover and prevents excessive accumulation. (**c**) We used the con_nanoLuc and test_nanoLuc mRNA to determine the effect of 5′-UTR structure on translation. The templates were amplified from gBlocks using primers FP_1 and RP_1. (**d**) The con_nanoLuc 5′-UTR is predicted to form a weak secondary structure with a free energy of ΔG −1.73 kcal/mol. (**e**) The test_nanoLuc 5′-UTR is predicted to form a stable stem-loop structure with a free energy of ΔG −15.60 kcal/mol

14. In vitro Transcription HiScribe T7 mRNA Kit with Clean Cap reagent.
15. DNase I.
16. Monarch RNA Cleanup Kit.
17. *E. coli* Poly(A) Polymerase.
18. Lipofectamine MessengerMAX Reagent.
19. 96-well cell culture plates.
20. SKNAS neuroblastoma cell line (ATCC).

Table 1
Primer sequences

Primer	Sequence
naLuc_repo-FP_1	5′-GCT AGC ACT ACC AGT CGC CAC C-3′
naLuc_repo-RP_1	5′-GGC TCG AGC GAT CGC CTA GGA A-3′
T7fireLucPCR FP_2	5′-GCT AGC ACT ACC AGT CGC CAC CTA ATA CGA CTC ACT ATA GAG AGA CAA GGA TAA AAG CAA GGA ATT CAG GTA GTG AAG AAA AAT CGA ACA CCA TGG CCG ATG CTA AGA ACA TTA AGA AGG
T7fireLucPCR RP_2	5′-GGC TCG AGC GAT CGC CTA GGA ACG CTA GAG TCG CGG CCT TAT TAC ACG GCG ATC TTG CCG CCT TTC

Underlined: T7 RNA polymerase promotor Double underlined: start codon or stop codon

21. Neubauer counting chamber.
22. ZOE Fluorescent Cell Imager or similar.
23. Clariostar or similar multiplate reader.
24. Dual-Glo Luciferase Assay System.
25. Plastic container for gel staining.
26. Low range single stranded RNA ladder (50–1000 nt).
27. Fluorescence imager.
28. 1.5 mL Eppendorf tubes.
29. Falcon tubes.
30. Heat block/mixer (thermo mixer).
31. Gel-loading tips.
32. Refrigerated tabletop centrifuge.
33. GraphPad Prism, a statistical graphing program.
34. Englisch spelling was optimized with DeepL.com (free version).

2.1 Solutions and Buffers

1. 20× MOPS: 400 mM MOPS, 100 mM $C_2H_3NaO_2$, 20 mM EDTA, pH 7.
2. 2× RNA Loading Dye: 95% Formamide, 0.02% SDS, 0.02% bromophenol blue, 0.01% Xylene Cyanol, 1 mM EDTA.
3. 6× Purple RNA Loading Dye: 15% Ficoll®-400, 60 mM EDTA, 19.8 mM Tris-HCl pH 8, 0.48% SDS, 0.12% Dye 1, 0.006% Dye 2.
4. TE Buffer: 10 mM Tris-HCl pH 7.5, 0.1 mM EDTA.
5. 10× TAE Buffer: 500 mM Tris-HCl pH7.4, 200 mM Sodium acetate, 20 mM EDTA.

3 Methods

3.1 gBLOCK Design

As shown, gBlocks contain different elements (see Fig. 1):

a. The T7 polymerase promoter sequence: 5′ TAATACGACT CACTATA**AG**. The bolded sequence AG is the transcriptional initiation sequence (*see* **Notes 3 and 4**).

b. 5′-UTR: Always consider a control sequence that is similar in length. In our example, the control sequence is similar in length but not structured (*see* **Note 5**).

c. Reporter ORF, we use a destabilized green fluorescence protein for establishing the expression kinetics and the transfection efficiency (see Fig. 2).

d. To monitor the impact of different 5′-UTRs on protein synthesis, we use the destabilized nanoLuc protein mRNA reporter. As the cotransfected normalization control, we use the mRNA reporter expressing the firefly luciferase protein (see Fig. 1).

3.2 PCR-Based Amplification of gBlocks

gBlocks are delivered as 1 μg pellets. To amplify the gBlocks, the gBlocks were resuspended in 100 μL TE Buffer. The PCR primers are shown in (see Table 1). Following PCR, the products are purified and are used as template for the generation of synthetic mRNAs (see Sect. 3.3).

1. Pipette the components according to the Table 2 into PCR cups.
2. Mix shortly and spin down.
3. Place the PCR tubes into an appropriate PCR machine.
4. Start the program (see Table 3).
5. Purify the PCR products after the run following the instructions on the PCR product purification kit (*see* **Note 6**).
6. Determine the amount of DNA in the eluate with Nanodrop 2000 or similar instrument (*see* **Note 7**).
7. Check DNA products for size and homogeneity by agarose gel electrophoresis.
8. Adjust the DNA amount to 100 ng/μL.

3.3 Generation of the DNA Template for the In Vitro Transcription of the fLuc mRNA

To generate a DNA template applicable for the in vitro transcription of fLuc mRNA, we used the pmirGLO Vector containing the fLUC open reading frame as template for PCR. The forward primer, T7fireLucPCR FP, contains elements necessary for in vitro transcription (*see* **Note 2**). The PCR products are purified and used as templates for generating synthetic fLuc mRNA (see Sect. 3.4).

1. Pipette the components according to the Table 4 into PCR cups.
2. Mix shortly and spin down.
3. Place the PCR tubes into an appropriate PCR machine.

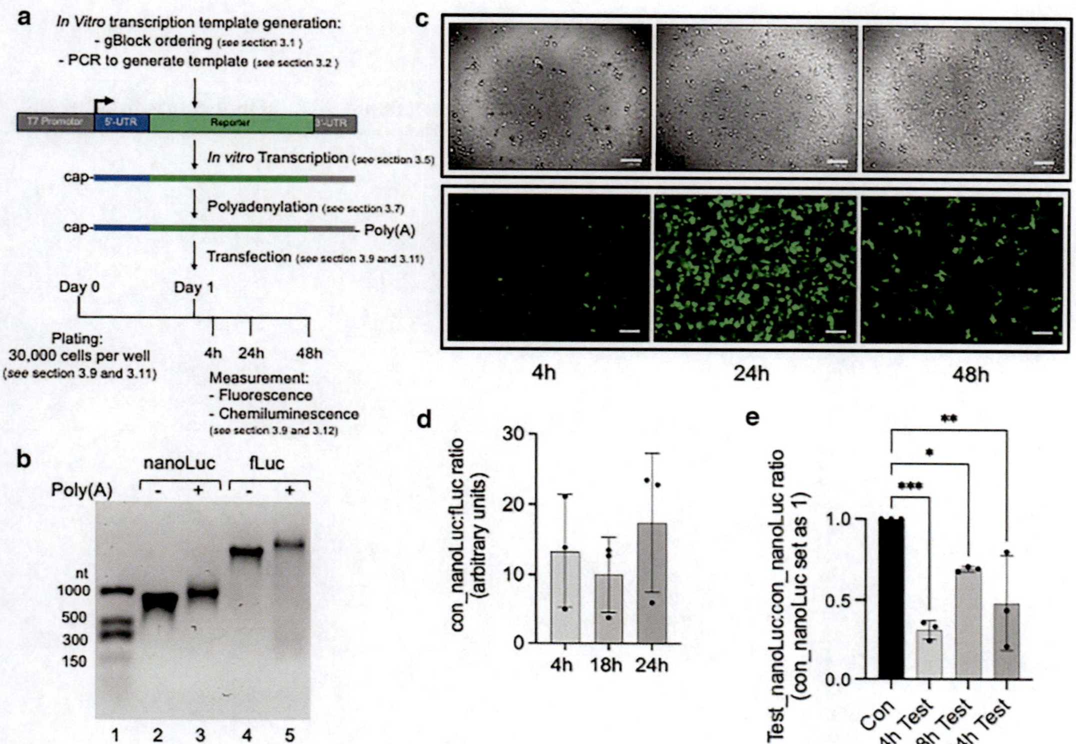

Fig. 2 Cell-based analysis showed a significant effect of the structure of the 5′UTR on translation. (**a**) This figure shows the different steps of the method for synthesizing capped and polyadenylated mRNA, as well as the time course of cell-based assays. (**b**) Denaturing RNA agarose gel showing capped and capped/polyadenylated con_nanoLuc and fLuc mRNA. As expected, polyadenylation increased the size of the mRNAs. Line 1= RNA Ladder; Line 2 = capped con_nanoLuc without poly(A) tail. Line 3 = capped con_nanoLuc with poly(A) tail. Line 4 = capped fLuc without poly(A) tail. Line 5 = capped fLuc with poly(A) tail. (**c**) mRNA transfection of SK-N-AS cells. The upper panel shows phase contrast images of SK-N-AS cells plated at a density of 30,000 cells/well. Phase contrast and fluorescence images of the same area were taken 4, 24, and 48 h after mRNA transfection. The lower panel shows that transfecting cells with con_destaGFP mRNA resulted in 70% confluency and detectable GFP expression at 4 h. By 24 h, the cells had reached approximately 100% confluency, with about 70% exhibiting GFP fluorescence. By 48 h, confluency decreased slightly to approximately 90%, with only about 15% of the cells retaining the GFP signal. (Scale bar = 100 μm). (**d**) Analysis of con_nanoLuc expression in SK-N-AS cells over time. The con_nanoLuc:fLuc ratio at 4, 18, and 24 h after transfection is shown. $n = 3$, ordinary one-way ANOVA test, multiple comparison. (**e**) A comparison of con_nanoLuc and test_nanoLuc expression shows significantly reduced expression of the stem-loop containing test_nanoLuc mRNA. The con_nanoLuc expression is set to 1, and the test_nanoLuc:con_nanoLuc ratios are shown (all values are fLuc-normalized). $n = 3$, ordinary one-way ANOVA test, multiple comparison, $p<0.05$

4. Start the program (see Table 5).
5. After the run, purify PCR products purification kit instruction (*see* **Note 6**).
6. Determine the amount of DNA in the eluate with Nanodrop 2000 or similar instrument.

Table 2
Pipetting scheme 1 for the PCR reaction 1

Component	Concentration	Volume	Final concentration
Q5 reaction buffer	5×	10 µL	1×
DNA template: gBlocks	10 ng/µL	1 µL	0.2 ng/µL
dNTPS	10 mM	1 µL	200 µM
naLuc_repo-FP	10 µM	2.5 µL	0.5 µM
naLuc_repo-RP	10 µM	2.5 µL	0.5 µM
Q5 high fidelity polymerase		0.5 µL	0.02 U/µL
$H_2$0 (nuclease-free)		32.5 µL	

Table 3
PCR Program 1

PCR Program 1

Temperature	Time	
98°C	30s	
98°C	10s	x25
54°C	30s	
72°C	30s	
72°C	2 min	
10°C	∞	

Table 4
Pipetting scheme 2 for the PCR reaction 2

Component	Concentration	Volume	Final concentration
Q5 Reaction Buffer	5×	10 µL	1×
DNA template:pmirGLO Vector	10 ng/µL	1 µL	0.2 ng/µL
dNTPS	10 mM	1 µL	200 µM
T7fireLucPCR FP	10 µM	2.5 µL	0.5 µM
T7fireLucPCR RP	10 µM	2.5 µL	0.5 µM
Q5 high fidelity polymerase		0.5 µL	0.02 U/µL
$H_2$0 (nuclease-free)		32.5 µL	

Table 5
PCR Program 2

PCR Program 2

Temperature	Time	
98°C	30s	
98°C	10s	
54°C	30s	x25
72°C	45s	x25
72°C	3 min	x25
10°C	-	

Table 6
Pipetting scheme for the in vitro transcription

Reagent	Concentration	Volume	Final concentration
T7 Cleancap Reagent AG reaction buffer	10×	2 µL	1×
ATP	60 mM	2 µL	6 mM
GTP	50 mM	2 µL	5 mM
UTP	50 mM	2 µL	5 mM
CTP	50 mM	2 µL	5 mM
CleanCap Reagent AG	40 mM	2 µL	4 mM
Template DNA	100 ng/µL	5 µL	0.5 µg
DTT	0.1 M	1 µL	5 µM
T7 RNA polymerase mix		2 µL	
Reaction volume		**20 µL**	

7. Check DNA product for size (expected size: 1786 nts) and homogeneity by 1% agarose gel electrophoresis.
8. Adjust the DNA amount to 100 ng/µL (Table 6).

3.4 Agarose Gel Electrophoreses to Evaluate the PCR Products

To ensure that the PCR products are the correct size and that only one product was generated, perform a 1% agarose gel electrophoresis to evaluate the success of your PCR.

1. Add 0.7 g Agarose to 70 mL of 1× TAE Buffer.
2. Heat the solution in a microwave oven to melt the agarose (*see* **Note 8**).

3. Add 3 μL Gel Red.
4. Swirl the gel solution to mix all components.
5. Pour the gel into the cast, insert appropriate comb.
6. Prepare DNA samples by mixing 500 ng of DNA with 6× purple loading dye, dilute reaction to 1× loading dye.
7. After the gel is fully cured, place into gel chamber with 1× TAE and load DNA ladder and DNA samples.
8. Run the gel at 100 V, 350 mA for 60 min.
9. Evaluate gel using a gel imager and the gel red filter setting.

3.5 In Vitro Transcription (IVT) to Generate the mRNA Reporter

In this section, the DNA templates (see Sects. 3.2 and 3.3) are used for in vitro transcription (IVT) with T7 RNA polymerase. During this step, a cap is added to the 5′-end (*see* **Note 3**). A polyA tail is added in a separate reaction to create the full synthetic mRNA (see Sect. 3.8 and Fig. 2a). We used the In vitro Transcription High scribe T7 mRNA Kit with Clean Cap reagent kit from NEB.

The following templates are used in this protocol:

a. con_destaGFP.
b. con_nanoLuc.
c. test_nanoLuc.
d. fLuc.

Slowly thaw the reagent, shortly mix and spin the tubes and keep them at RT. Keep the enzyme mix at 4°C.

1. Pipet the reaction according to the Table 7. Mix shortly and spin down.
2. Incubate for 2 h at 37°C (*see* **Note 9**).
3. Continue with Sect. 3.7.

3.6 Removal of the DNA Template by DNase I Digestion

To remove the DNA templates from the in vitro transcription reaction, a DNase I digestion was performed.

1. Add 30 μL nuclease-free water to the 20 μL IVT reaction (see Sect. 3.6) reaching a total volume of 50 μL.

Table 7
Pipetting scheme for the Poly(A) tailing reaction

Reagent	Concentration or amount	Volume
RNA	10 μg	In 15 μL nuclease free H_2O
E. coli Poly(A) polymerase reaction buffer	10×	2 μL
ATP	10 mM	2 μL
E. coli Poly (A) polymerase		1 μL
Reaction volume		**20 μL**

2. Add 2 µL DNase I.
3. Shortly mix and spin down.
4. Incubate reaction for 15 min at 37°C.
5. For purification of the RNA, we are using the Monarch RNA Cleanup Kit.
6. Therefore, add 100 µL Monarch Buffer BX to the 50 µL IVT reaction.
7. Add 1 Vol. (150 µL) Ethanol.
8. Mix well by pipetting up and down.
9. Place the column into the collection tube.
10. Add the reaction (300 µL) onto the column and close the lid.
11. Spin for 1 min (16,000 × *g*).
12. Discard flowthrough.
13. Transfer the column to a new collection tube.
14. Add 500 µL of Monarch Puffer WX.
15. Spin for 1 min (16,000 × *g*).
16. Discard flowthrough.
17. Repeat Steps 14–16.
18. Place the column in a nuclease-free 1.5 mL Eppendorf tube.
19. For RNA elution, add 33 µL nuclease-free water.
20. Spin for 1 min (16,000 × *g*).
21. Mix the eluate and spin down.
22. Determine the concentration using a Nanodrop or similar device (*see* **Note 10**).

3.7 Poly(A) Tail Synthesis by E. coli Poly(A) Polymerase

In the final step, the capped purified synthetic in vitro transcripts are used for poly(A) tailing (*see* **Note 11**).

The following in vitro transcribed mRNAs are used:

a. Con_destaGFP mRNA.
b. Con_nanoLuc mRNA.
c. Test_nanoLuc mRNA
d. fLuc mRNA.

1. You need 10 µg RNA per poly(A) tailing reaction.
2. **Important:** Include a reaction without poly(A) polymerase to judge the success of the polyA tailing reaction in agarose gel electrophoresis.
3. Add the components according to Table 8 into PCR cups.
4. Mix carefully and spin down.
5. Incubate reactions for 60 min at 37°C (*see* **Note 11**).

Table 8
1% Denaturierende agarose formaldehyde gel

Component	Amount
Agarose	1 g
H_2O	77 mL
20× MOPS	5 mL
Gel-Red	6 μL
37% Formaldehyde	18 mL

6. Finally, the poly(A) RNA is purified again by using the "Monarch RNA Cleanup Kit" (see Sect. 3.6 Steps 6–21).
7. Determine the concentration using a Nanodrop or similar device.
8. **RNA storage**: adjust the RNA to 200 ng/μL with nuclease-free water. Prepare aliquots and store the RNA at 80°C (*see* **Note 12**).

3.8 Denaturing Agarose Gel Electrophoresis to Check for Synthetic mRNA and poly(A) Tailing

Due to the length of the synthetic RNA, a denaturing agarose gel is recommended for checking the homogeneity and length of the poly(A) tail (see Fig. 2b, *see* **Note 13**).

1. Add 1 g Agarose to 77 mL nuclease free water.
2. Heat the solution in a microwave oven to melt the agarose (*see* **Note 8**).
3. Add 5 mL 20× MOPS.
4. Add 6 μL Gel-Red.
5. Swirl the gel solution to mix all components.
6. Measure temperature of the solution with an infrared thermometer.
7. At 50–55°C add 18 mL 37% Formaldehyde. Important safety issue: pipette under the hood (*see* **Note 13**).
8. Pour the gel into the cast. Important safety issue: perform this step under a hood.
9. Prepare probes by mixing 500 ng RNA in 5 μL H_20 plus 5 μL 2× RNA Loading Dye.
10. Incubate the reaction at 70°C for 10 min.
11. Fill the gel chamber with 1× MOPS.
12. Load the sample.
13. Load ladder (*see* **Note 14**).
14. Run the gel at 80 V and 350 mA for 80 min.
15. Take a picture of the gel red fluorescence signal in the gel and label lanes (see Fig. 2).

3.9 RNA Transfection

In our case, we are using the neuroblastoma cell line SK-N-AS. The prepared and characterized synthetic mRNAs are transfected into the cell using the Lipofectamine MessengerMAX Reagent (see Fig. 2a). First, we transfected the con_destaGFP mRNA to test the transfection efficiency and the expression over 4, 24, and 48 h (see Fig. 2c).

1. **Day 0:** Continuous SK-N-AS cell cultures were split every 3 days, and cells at a confluence of no more than 80% were harvested by trypsinization and counted for plating into 96-well plates. 30,000 cells in 100 μL cell culture medium were plated per well of a 96-well plate (*see* **Note 15**). Use clear 96-well cell culture plates when studying the GFP expression and complete white 96-well cell culture plates when measuring luciferase expression (*see* **Note 16**).
2. **Day 1:** Transfection were always prepared in triplicates.
3. Pipet reaction according to Table 9.
4. Mix very gently (*see* **Note 17**).
5. Incubate 10 min at RT.
6. Pipet reaction according to Table 10.
7. Mix gently (*see* **Note 17**).
8. Combine transfection reaction 1 and 2.
9. Mix very gently by pipetting up and down (*see* **Note 17**).
10. Incubate 5 min at RT.
11. Distribute transfection reaction by pipetting 10 μL/well.
12. Incubate the plate in the cell culture incubator.
13. At 4, 24, and 48 h after transfection take phase contrast and fluorescence pictures with the ZOE imager or any other fluorescence microscope (see Fig. 2b).

Table 9
Transfection reaction 1

	1×	2×
Opti-MeM	5 μL	15 μL
MessengerMax reagent	0.3 μL	0.9 μL

Table 10
Transfection reaction 2

	1×	3×
Opti-MeM	5 μL	15 μL
Con_destaGFP (200 ng/μL)	0.5 μL	1.5 μL

14. After 4 h, we already observed a clear GFP signal. The strongest signal was observed at 24 h and a weaker signal at 48 h, suggesting that the decay of the synthetic mRNAs took place.

3.10 RNA Cotransfection

Based on the results of the pilot experiment using the con_destaGFP mRNA, we decided to cotransfect (*see* **Note 18**) the con_nanoLuc, the test_nanoLuc mRNAs with the fLuc mRNA for normalization. We designed the experiments so that we would have transfected cell culture plates for analysis at 4, 18, and 24 h after transfection (see Fig. 2d and e, *see* **Note 19**). Please note that three wells containing cells that have not been transfected are required as a blank value for the luciferase assay.

1. **Day 0:** Continuous SK-N-AS cell cultures were split every 3 days, and cells at a confluence of no more than 80% were harvested by trypsinization and counted for plating into white 96-well plates, which were used when measuring luciferase expression (*see* **Note 19**).
2. 30,000 cells (*see* **Note 15**) in 100 μL cell culture medium were plated per well of a 96-well plate.
3. **Day 1**: Transfections were always prepared in triplicates.
4. Mix reaction according to Table 9.
5. Incubate for 10 min at RT.
6. Pipet reaction according to Table 11.
7. Mix gently (*see* **Note 17**).
8. Combine transfection reaction 1 and 2.
9. Mix very gently by pipetting up and down (*see* **Note 17**).
10. Incubate 5 min at RT.
11. Distribute transfection reaction by pipetting 10 μL/well.
12. Incubate the plate in the cell culture incubator.

Table 11
Transfection reaction con_nanoLuc/fLuc

	1×	3×
Opti-MeM	5 μL	15 μL
Con_nanoluc mRNA (200 ng/μL)	0.25 μL	0.75 μL
fLuc mRNA (200 ng/μL)	0.25 μL	0.75 μL
Transfection reaction test_nanoLuc/fLuc:		
	1×	3×
Opti-MeM	5 μL	15 μL
test_nanoluc mRNA (200 ng/μL)	0.25 μL	0.75 μL
fLuc mRNA (200 ng/μL)	0.25 μL	0.75 μL

13. At 4 h, after transfection measure luciferase activity (see Sect. 3.10).
14. **Day 2:** determine luciferase expression 18 and 24 h after transfection (see Sect. 3.12).

3.11 Luciferase Measurement

When measuring luciferase, remove the 96-well plate containing the transfected cells from the incubator, then proceed as shown below. It is important to note that you must first measure the Firefly luciferase, followed by the second nanoLuc measurement. Therefore, it is critical that you use the Firefly Luciferase setting on the plate reader and add the Dual-Glo Luciferase reagent first. For the second nanoLuc measurement, change the plate reader settings and use the Dual-Glo Stop & Glo Reagent (*see* **Note 20**). Mixing up the order will render the data unusable.

Chemiluminescence measurement:

1. Use a multichannel pipette to remove 80 μL of the cell culture media.
2. Add 30 μL Dual-Glo Luciferase Reagent (*see* **Note 21**).
3. Mix the plate at a rotation shaker for 10 min at RT.
4. Measure Firefly Luciferase chemiluminescence first (*see* **Note 20**).
5. Add 30 μL Dual-Glo Stop & Glo Reagent per well (*see* **Note 22**).
6. Mix the plate at a rotation shaker for 10 min at RT.
7. Measure nanoLuc chemiluminescence (*see* **Note 20**).

Program settings for the Clariostar plate reader:

Basic settings

Measurement type: Luminescence
Microplate name: GREINER 96 F-BOTTOM

Well mode settings

No. of intervals: 1
Measurement interval time [s]: 1.77
Scan mode: spiral scan
Scan diameter [mm]: 6

Optic settings

Presetname: Firefly
Emission: 580-80
Gain: 3600
Well(s) used for gain adjustment: B4
Focal height [mm]: 12.0

Shaking settings

Shaking frequency [rpm]: 400
Shaking mode: double orbital
Additional shaking time: 2s before plate reading

General settings

Top optic used
Aperture spoon: -
Injection needle holder type: -
Settling time [s]: 0.2
Reading direction:

Target temperature [°C]: set off
Target concentration O2 [%]: set off
Target concentration CO2 [%]: set off

To allow comparison of measurements with different measurement interval times values are normalized to 0.1 sec.

For firefly luciferase measurement.

Basic settings

Measurement type: Luminescence
Microplate name: GREINER 96 F-BOTTOM

Well mode settings

No. of intervals:	1	Scan mode:	spiral scan
Measurement interval time [s]:	1.77	Scan diameter [mm]:	6

Optic settings

Presetname:	NanoLuc
Emission:	470-80
Gain:	3600
Well(s) used for gain adjustment:	A3
Focal height [mm]:	9.0

Shaking settings

Shaking frequency [rpm]:	400
Shaking mode:	double orbital
Additional shaking time:	2s before plate reading

General settings

Top optic used	
Aperture spoon:	-
Injection needle holder type:	-
Settling time [s]:	0.2
Reading direction:	
Target temperature [°C]:	set off
Target concentration O2 [%]:	set off
Target concentration CO2 [%]:	set off

To allow comparison of measurements with different measurement interval times values are normalized to 0.1 sec.

For nanoLuc luciferase measurement.

3.12 Analysis

The goal of the protocol was to develop an experimental procedure for studying the control of protein synthesis by elements in the 5′ -UTR of the reporter mRNA. To this end, we compared the translation of control and test reporter mRNA that differed only in their 5′-UTR features. After completing the measurement of both Nano Luciferase and Firefly Luciferase chemiluminescence, we took the blank (cells that were not transfected) corrected values of nanoLuc and calculated the mean of each triplicate. The same was done for the fLuci values. We then took the nanoLuc mean and divided it by the fLuc mean of the same triplicate to account for differences in transfection efficiency. This gave us the blank corrected and normalized nanoLuc value of each triplicate. From here, we divided the nanoLuc values of the test_nanoLuc mRNa transfected cells by the nanoLuc values of the con_nanoLuc mRNA transfected cells. This gave us the ratio of the translation test_nanoLuc mRNA in comparison to con_nanoLuc mRNA. A ratio greater than one would indicate that the test_nanoLuc mRNA was translated more, whereas a ratio less than one would indicate lower translation. Using our method, we detected that the stem loop that is in the 5′UTR of test_nanoLuc mRNA (Fig. 1) has a significant inhibitory effect on mRNA translation (Fig. 2e). The translational repression was statistically significant at 4, 18, and 24 h ($p<0.05$), confirming that the stem-loop in the 5′UTR markedly impairs translation of the test mRNA.

4 Notes

1. Ribonucleases are present on the skin and quickly contaminate all equipment and pipettes. Therefore, special precautions are required when working with RNA. Always clean the workbench, pipettes, and racks with 70% ethanol. Wear gloves and change them frequently or wipe them with 70% ethanol towels. Use RNase-free water and solutions. Furthermore, adhere to institutional safety and reagent disposal rules.
2. The T7fireLucPCR_FP_2 contains a T7 promoter sequence (underlined), a control 5′-UTR, and the start codon (AUG, double underlined) for the Firefly Luciferase (fLuc) open reading frame.
3. The T7 RNA polymerase promoter contains an AG (in bold, see Fig. 1) as the transcriptional initiation sequence. This sequence is necessary for an efficient capping step when using NEB CleanCap reagents.
4. When planning to clone gBlocks into a vector, consider adding a sequence containing sites for appropriate restriction enzymes to the 5′- and 3′-ends of the gBlock.
5. We use the ViennaRNA Web Services (http://rna.tbi.univie.ac.at) for RNA folding predictions of RNA sequences.
6. The purity of the DNA templates is critical for efficient in vitro transcription. If you gel-purify the PCR product, be sure to perform the additional purification/wash step, which is often listed as optional.
7. We obtained PCR product yields of around 150 ng/μL.
8. Be careful when removing the heated solution from the microwave. The movement created in the liquid can cause delayed boiling, which can result in hot liquid shooting out of the beaker. Always angle the beaker opening away from your body and face.
9. We used a PCR instrument set to 37°C (with a heated lid) to prevent evaporation. For reactions with transcripts <0.3 kb incubations up to 16 h or overnight can be performed.
10. The RNA yield for templates that do not encode a stem-loop RNA structure is usually between 1400 and 2200 ng/μL. Templates that encoded a stem-loop structure yielded an average of around 500 ng/μL.

11. This is a guideline for the polyA tailing reaction. Incubate 5 units of the enzyme with 1–10 µg of RNA in a 20 µL reaction at 37°C for 30 min with 1× reaction buffer and 1 mM ATP to achieve a tail length greater than 100 bases.
12. For our experiments, we found that an aliquot volume of 20 µ L was reasonable. The RNA was diluted to 200 ng/µL with nuclease-free water. However, you should adjust the volume of the aliquot to the size of your planned experiments. We used a RNA ladder from 50–1000 nucleotides.
13. Formaldehyde is toxic, carcinogenic, and a strong irritant, and must be handled in a well-ventilated area or fume hood while wearing appropriate protective equipment.
14. We used a RNA ladder from 50–1000 nucleotides.
15. We achieved appropriate confluency (70%) at day of transfection by plating 30,000 SK-N-AS cells per well (96-well plate).
16. Be aware that no cells are visible and no light microscopy is possible when using white plates without a clear bottom.
17. Carefully mix the transfection reactions to prevent the transfection reagent from spreading to the walls of the tube.
18. For cotransfection, it is important to determine the optimal ratio of the two different mRNAs. Keep in mind that both mRNAs compete for cellular translation machinery, and the length of the open reading frames may differ. This means that a short protein is quickly produced, while a long protein takes longer to reach a measurable level.
19. To determine luciferase expression at different time points, we used one plate for each time point because the luciferase assay destroys the cells, and shaking and incubating the plates at room temperature would harm the other cells on the plates. Each plate contained technical triplicates. Furthermore, it is recommended to only work with one cell line per plate to avoid cross-contamination.
20. Firefly and Nano luciferases emit light at different wavelengths due to different bioluminescent chemical reactions. The Firefly Luciferase emission peak is around 560 nm, while the Nano luciferase emission peak is around 470–480 nm. Remember to change the settings!
21. Thaw the luciferase reagents on ice.
22. Dilute the yellow Dual-Glo Stop & Glo Reagent at a ratio of 1:100 using the provided Dual-Glo Stop & Glo buffer solution. It is important to prepare the Dual-Glo Stop & Glo Reagent solution in a small glass beaker. Do not use plastic tubes.

Acknowledgments

We would like to thank Marie Matthes for her technical support, as well as the Verein zur Förderung krebskranker und körperbehinderter Kinder Ostbayern e.V. (VKKK), for their support. Financial support for this work was provided by the Winkler Foundation (Parkstein, Germany).

References

1. Hinnebusch AG, Ivanov IP, Sonenberg N (2016) Translational control by 5′-untranslated regions of eukaryotic mRNAs. Science 352(6292):1413–1416. https://doi.org/10.1126/science.aad9868
2. Gebauer F, Hentze MW (2004) Molecular mechanisms of translational control. Nat Rev Mol Cell Biol 5(10):827–835
3. Silvera D, Formenti SC, Schneider RJ (2010) Translational control in cancer. Nat Rev Cancer 10(4):254–266. https://doi.org/10.1038/nrc2824
4. Kozak M (1987) An analysis of 5′-noncoding sequences from 699 vertebrate messenger RNAs. Nucleic Acids Res 15(20):8125–8148
5. Mahe M, Rios-Fuller T, Katsara O, Schneider RJ (2024) Non-canonical mRNA translation initiation in cell stress and cancer. NAR Cancer 6(2):zcae026. https://doi.org/10.1093/narcan/zcae026
6. Sriram A, Bohlen J, Teleman AA (2018) Translation acrobatics: how cancer cells exploit alternate modes of translational initiation. EMBO Rep 19(10):e45947. https://doi.org/10.15252/embr.201845947
7. Wek RC (2018) Role of eIF2alpha kinases in translational control and adaptation to cellular stress. Cold Spring Harb Perspect Biol 10(7):a032870. https://doi.org/10.1101/cshperspect.a032870
8. Hinnebusch AG (1993) Gene-specific translational control of the yeast GCN4 gene by phosphorylation of eukaryotic initiation factor 2. Mol Microbiol 10(2):215–223. https://doi.org/10.1111/j.1365-2958.1993.tb01947.x
9. Richter JD, Sonenberg N (2005) Regulation of cap-dependent translation by eIF4E inhibitory proteins. Nature 433(7025):477–480
10. Proud CG (2002) Regulation of mammalian translation factors by nutrients. Eur J Biochem 269(22):5338–5349
11. Haghighat A, Mader S, Pause A, Sonenberg N (1995) Repression of cap-dependent translation by 4E-binding protein 1: competition with p220 for binding to eukaryotic initiation factor-4E. EMBO J 14(22):5701–5709. https://doi.org/10.1002/j.1460-2075.1995.tb00257.x
12. Pause A, Belsham GJ, Gingras AC, Donze O, Lin TA, Lawrence JC Jr, Sonenberg N (1994) Insulin-dependent stimulation of protein synthesis by phosphorylation of a regulator of 5′-cap function. Nature 371(6500):762–767. https://doi.org/10.1038/371762a0
13. Truitt ML, Ruggero D (2016) New frontiers in translational control of the cancer genome. Nat Rev Cancer 16(5):288–304. https://doi.org/10.1038/nrc.2016.27
14. Leppek K, Das R, Barna M (2018) Functional 5′ UTR mRNA structures in eukaryotic translation regulation and how to find them. Nat Rev Mol Cell Biol 19(3):158–174. https://doi.org/10.1038/nrm.2017.103
15. Godet AC, David F, Hantelys F, Tatin F, Lacazette E, Garmy-Susini B, Prats AC (2019) IRES trans-acting factors, key actors of the stress response. Int J Mol Sci 20(4):924. https://doi.org/10.3390/ijms20040924
16. Pyronnet S, Pradayrol L, Sonenberg N (2000) A cell cycle-dependent internal ribosome entry site. Mol Cell 5(4):607–616
17. Johannes G, Carter MS, Eisen MB, Brown PO, Sarnow P (1999) Identification of eukaryotic mRNAs that are translated at reduced cap binding complex eIF4F concentrations using a cDNA microarray. Proc Natl Acad Sci U S A 96(23):13118–13123
18. Macejak DG, Sarnow P (1991) Internal initiation of translation mediated by the 5′ leader of a cellular mRNA. Nature 353(6339):90–94. https://doi.org/10.1038/353090a0
19. Hellen CU, Pestova TV (1999) Translation of hepatitis C virus RNA. J Viral Hepatitis 6(2):79–87. https://doi.org/10.1046/j.1365-2893.1999.00150.x
20. Hellen CU, Sarnow P (2001) Internal ribosome entry sites in eukaryotic mRNA molecules. Genes Dev 15(13):1593–1612. https://doi.org/10.1101/gad.891101

21. Ryabova LA, Pooggin MM, Hohn T (2002) Viral strategies of translation initiation: ribosomal shunt and reinitiation. Prog Nucleic Acid Res Mol Biol 72:1–39. https://doi.org/10.1016/s0079-6603(02)72066-7
22. Lee AS, Kranzusch PJ, Cate JH (2015) eIF3 targets cell-proliferation messenger RNAs for translational activation or repression. Nature 522(7554):111–114. https://doi.org/10.1038/nature14267
23. Dever TE, Ivanov IP, Hinnebusch AG (2023) Translational regulation by uORFs and start codon selection stringency. Genes Dev 37(11-12):474–489. https://doi.org/10.1101/gad.350752.123
24. Avni D, Biberman Y, Meyuhas O (1997) The 5′ terminal oligopyrimidine tract confers translational control on TOP mRNAs in a cell type- and sequence context-dependent manner. Nucleic Acids Res 25(5):995–1001
25. Ledda M, Di Croce M, Bedini B, Wannenes F, Corvaro M, Boyl PP, Caldarola S, Loreni F, Amaldi F (2005) Effect of 3′UTR length on the translational regulation of 5′-terminal oligopyrimidine mRNAs. Gene 344:213–220. https://doi.org/10.1016/j.gene.2004.09.023 S0378-1119(04)00597-9 [pii]
26. Fonseca BD, Zakaria C, Jia JJ, Graber TE, Svitkin Y, Tahmasebi S, Healy D, Hoang HD, Jensen JM, Diao IT, Lussier A, Dajadian C, Padmanabhan N, Wang W, Matta-Camacho E, Hearnden J, Smith EM, Tsukumo Y, Yanagiya A, Morita M, Petroulakis E, Gonzalez JL, Hernandez G, Alain T, Damgaard CK (2015) La-related Protein 1 (LARP1) represses terminal oligopyrimidine (TOP) mRNA translation downstream of mTOR complex 1 (mTORC1. J Biol Chem 290(26):15996–16020. https://doi.org/10.1074/jbc.M114.621730
27. Tcherkezian J, Cargnello M, Romeo Y, Huttlin EL, Lavoie G, Gygi SP, Roux PP (2014) Proteomic analysis of cap-dependent translation identifies LARP1 as a key regulator of 5′TOP mRNA translation. Genes Dev 28(4):357–371. https://doi.org/10.1101/gad.231407.113
28. Damgaard CK, Lykke-Andersen J (2011) Translational coregulation of 5′TOP mRNAs by TIA-1 and TIAR. Genes Dev 25(19):2057–2068. https://doi.org/10.1101/gad.17355911
29. Nandagopal N, Roux PP (2015) Regulation of global and specific mRNA translation by the mTOR signaling pathway. Translation (Austin) 3(1):e983402. https://doi.org/10.4161/21690731.2014.983402
30. Sen ND, Zhou F, Ingolia NT, Hinnebusch AG (2015) Genome-wide analysis of translational efficiency reveals distinct but overlapping functions of yeast DEAD-box RNA helicases Ded1 and eIF4A. Genome Res 25(8):1196–1205. https://doi.org/10.1101/gr.191601.115
31. Marsden S, Nardelli M, Linder P, McCarthy JE (2006) Unwinding single RNA molecules using helicases involved in eukaryotic translation initiation. J Mol Biol 361(2):327–335 https://doi.org/10.1016/j.jmb.2006.06.016
32. Parsyan A, Svitkin Y, Shahbazian D, Gkogkas C, Lasko P, Merrick WC, Sonenberg N (2011) mRNA helicases: the tacticians of translational control. Nat Rev Mol Cell Biol 12(4):235–245. https://doi.org/10.1038/nrm3083
33. Kuehnert J, Sommer G, Zierk AW, Fedarovich A, Brock A, Fedarovich D, Heise T (2015) Novel RNA chaperone domain of RNA-binding protein La is regulated by AKT phosphorylation. Nucleic Acids Res 43(1):581–594. https://doi.org/10.1093/nar/gku1309
34. Croitoru V, Semrad K, Prenninger S, Rajkowitsch L, Vejen M, Laursen BS, Sperling-Petersen HU, Isaksson LA (2006) RNA chaperone activity of translation initiation factor IF1. Biochimie 88(12):1875–1882. https://doi.org/10.1016/j.biochi.2006.06.017
35. Ishiguro T, Sato N, Ueyama M, Fujikake N, Sellier C, Kanegami A, Tokuda E, Zamiri B, Gall-Duncan T, Mirceta M, Furukawa Y, Yokota T, Wada K, Taylor JP, Pearson CE, Charlet-Berguerand N, Mizusawa H, Nagai Y, Ishikawa K (2017) Regulatory role of RNA chaperone TDP-43 for RNA misfolding and repeat-associated translation in SCA31. Neuron 94(1):108–124.e107. https://doi.org/10.1016/j.neuron.2017.02.046
36. Heise T, Kota V, Brock A, Morris AB, Rodriguez RM, Zierk AW, Howe PH, Sommer G (2016) The La protein counteracts cisplatin-induced cell death by stimulating protein synthesis of anti-apoptotic factor Bcl2. Oncotarget 7(20):29664–29676. https://doi.org/10.18632/oncotarget.8819
37. Truitt ML, Conn CS, Shi Z, Pang X, Tokuyasu T, Coady AM, Seo Y, Barna M, Ruggero D (2015) Differential requirements for eIF4E dose in normal development and cancer. Cell 162(1):59–71. https://doi.org/10.1016/j.cell.2015.05.049
38. Cunningham JT, Moreno MV, Lodi A, Ronen SM, Ruggero D (2014) Protein and nucleotide biosynthesis are coupled by a single rate-limiting enzyme, PRPS2, to drive cancer. Cell 157(5):1088–1103. https://doi.org/10.1016/j.cell.2014.03.052
39. Hsieh AC, Liu Y, Edlind MP, Ingolia NT, Janes MR, Sher A, Shi EY, Stumpf CR, Christensen C, Bonham MJ, Wang S, Ren P, Martin M, Jessen K, Feldman ME, Weissman JS, Shokat

KM, Rommel C, Ruggero D (2012) The translational landscape of mTOR signalling steers cancer initiation and metastasis. Nature 485(7396):55–61. https://doi.org/10.1038/nature10912

40. Xue S, Tian S, Fujii K, Kladwang W, Das R, Barna M (2015) RNA regulons in Hox 5′ UTRs confer ribosome specificity to gene regulation. Nature 517(7532):33–38. https://doi.org/10.1038/nature14010

Check for updates

Chapter 17

5EU-Based Tracking of RNA Transfer Through Extracellular Vesicles

Ji Won Lee, Chul Woong Ho and Kyung-Won Min

Abstract

Extracellular vesicles (EVs) are membrane-bound nanoparticles that mediate the intercellular transfer of diverse molecules. RNA species within EVs act as key regulators of gene expression and influence the phenotype in recipient cells. However, tracking RNA transfer via EVs remains technically challenging, as post-isolation fluorescent labeling can lead to artificial signals and genetic manipulation of recipient cells may disrupt endogenous regulatory pathways. Here, we detail a metabolic labeling approach with 5-ethynyl uridine (5-EU) that enables tracing of EV-mediated RNA transfer under physiological conditions. In this approach, donor cells incorporate 5-EU into nascent RNAs, which are naturally packaged into EVs. Following delivery to recipient cells, the transferred RNAs are selectively enriched via biotin-based click chemistry for downstream analysis. This method preserves native RNA loading into EVs and enables tracing their delivery to recipient cells, providing a reliable and minimally perturbative approach for studying EV-mediated RNA transport in intercellular communication.

Keywords Extracellular vesicles, Metabolic labeling, 5-ethynyl uridine (5-EU), RNA tracking

1 Introduction

Extracellular vesicles (EVs) are membrane-enclosed nanoparticles secreted by most cell types and play a pivotal roles in intercellular communication [1]. They carry diverse molecular cargoes, including proteins, lipids, and nucleic acids, that are delivered to recipient cells and modulate their phenotype and function [2, 3]. Among these cargoes, microRNAs (miRNAs) carried by EVs are well-recognized mediators of intercellular communication [4]. Once delivered, EV-miRNAs can modulate gene expression through post-transcriptional gene silencing, remodel signaling pathways, and influence immune responses, ultimately shaping the phenotype of recipient cells in both physiological and pathological contexts [5, 6].

While the presence of miRNAs within EVs is well established, the mechanisms regulating their selective loading remain unknown [7]. Increasing evidence has identified RNA-binding proteins (RBPs)

Je-Hyun Yoon (ed.), *RNA Labeling*, Methods in Molecular Biology, vol. 3047,
https://doi.org/10.1007/978-1-0716-5352-4_17,

as key regulators of miRNA sorting into EVs, including human antigen R (HuR) [8], major vault protein (MVP) [9], synaptotagmin-binding, cytoplasmic RNA-interacting protein (SYNCRIP)[10, 11], and poly(rC)-binding protein 2 (PCBP2) [11], among others. These RBPs interact with specific sequence motifs within miRNAs, or directly bind to target miRNAs, to promote or inhibit their EV loading [8–14]. Our previous study identified RBP quaking (QKI) as a suppressor of the EV-mediated release of the miRNA *let-7b* [15]. QKI binds to and retains mature *let-7b* in donor cells, preventing its release into EVs. This retention prevents toll-like receptor 7 (TLR7) signaling in recipient cells from being triggered by RNA cargo, thereby altering their immune phenotype. As such, RBPs not only shape the molecular composition of EVs, but also influence that miRNAs are ultimately transferred to recipient cells, thereby playing a critical role in modulating cell-to-cell communication.

Despite these mechanistic insights, it remains technically challenging to trace the transfer of miRNAs from donor cells to recipient cells via EVs. Current approaches, such as fluorescent labeling of RNA and genetic depletion of specific RNAs in recipient cells, suffer from limitations. Fluorescent RNA-binding dyes are commonly used to visualize EV-associated RNAs [16–19]. However, these probes are typically applied post-isolation, which fails to reflect the physiological packing of RNAs into EVs. In addition, they may stain surface-bound nucleic acids or induce vesicle/dye aggregation, which can confound downstream analysis and compromise the integrity of EVs [17]. Genetically engineered recipient models lacking specific miRNAs confirm the EV-mediated transfer by assessing the restoration of the target RNA following EV treatment [20]. However, such experimental manipulation may interfere with endogenous regulatory networks in recipient cells and complicate the interpretation of downstream results.

To address these challenges, we employ a metabolic labeling strategy using 5-ethynyl uridine (5-EU), a uridine analog that is metabolically incorporated into nascent RNAs in donor cells during transcription [15, 21, 22]. This endogenous labeling preserves the physiological process of RNA sorting into EVs. Following the uptake of EVs by recipient cells, transferred RNAs can be selectively enriched through biotin-based click chemistry and analyzed by RT-qPCR or RNA sequencing (Fig. 1). Although 5-EU is incorporated into all newly synthesized RNAs rather than targeting specific RNA species, it enables the comprehensive assessment of which RNAs have been transferred via EVs. We validated this method using *let-7b* miRNA under QKI-depleted conditions, demonstrating its utility in studying miRNA transfer via EVs [15]. This approach provides a valuable tool for tracing EV-mediated RNA transfer and elucidating the role of RNA in intercellular communication.

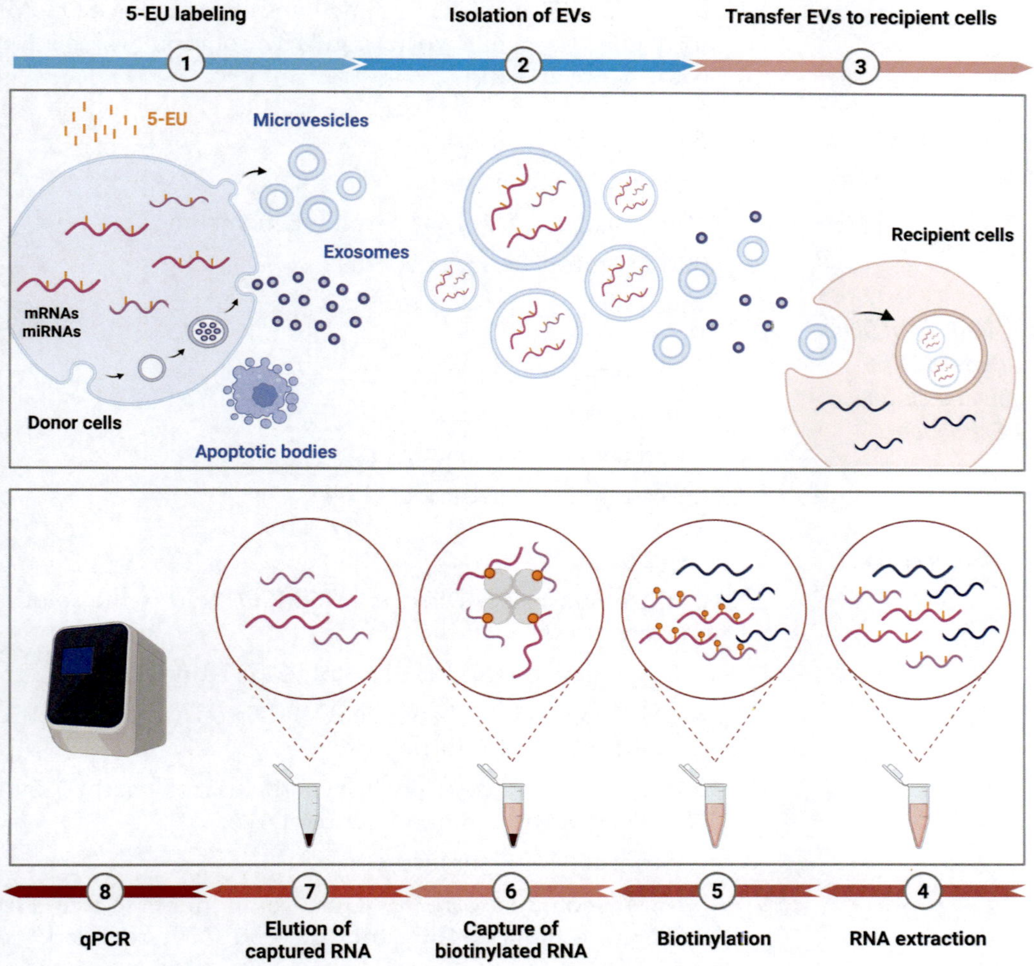

Fig. 1 Schematic diagram of tracking 5-EU labeled RNA transfer from donor cells to recipient cells via extracellular vesicles (EVs). The diagram depicts the experimental procedure for tracking RNAs derived from donor cells to recipient cells through EVs. Donor cells are metabolically labeled with 5-ethynyl uridine (5-EU), which is incorporated into nascent RNAs. EVs are isolated from the donor cells and applied to recipient cells. Following cellular uptake, 5-EU-labeled RNAs in recipient cells are biotinylated via click chemistry and enriched using streptavidin-coated beads. Transferred RNAs are quantified by RT-qPCR

2 Materials

2.1 5-EU Labeling of Nascent RNA

1. Donor cells (HeLa) and recipient cells (SH-SY5Y).
2. Dulbecco's Modified Eagle Medium (DMEM) (*see* **Note 1**).
3. 10% Fetal bovine serum (FBS).
4. 100 U/mL Penicillin/streptomycin.
5. Ice-cold Phosphate-Buffered Saline (PBS): 137 mM NaCl, 2.7 mM KCl, 10 mM Na_2HPO_4, and 1.5 mM KH_2PO_4.

6. 200 mM 5-EU: Dissolve in dimethyl sulfoxide (DMSO). Mix thoroughly by gentle vortexing. The solution will be dark brown. Store at −20°C.

2.2 Isolation of Extracellular Vesicles

1. PBS (see Sect. 2.1).
2. 1 U/mL of benzonase.
3. 0.2 μm asymmetric polyether sulfone membrane.
4. 500-kDa cutoff Biomax cassette.

2.3 Transfer of Donor Cell-Derived EVs to Recipient Cells and RNA Extraction

1. Trizol.
2. Chloroform.
3. Isopropanol.
4. 75% ethanol.
5. Nuclease-free distilled water (DW).

2.4 Biotinylation of RNA

1. 1–5 μg RNA.
2. 10 mM biotin azide: Dissolve in DMSO. Vortex the solution and store at −20°C in the dark.
3. 2× EU buffer: 100 mM HEPES pH 7.5, 10 mM EDTA.
4. 25 mM Copper (II) sulfate ($CuSO_4$): Dissolve in nuclease-free DW. Store at 4°C for further use.
5. 400 mM Tris(3-hydroxypropyltriazolylmethyl)amine (THPTA): Dissolve in nuclease-free DW.
   - Store at −20°C for further use.
6. 400 mM sodium ascorbate: Dissolve in nuclease-free DW. Divide into aliquots and store at −20°C in the dark (*see* **Note 2**).

2.5 Precipitation of Biotinylated RNA

1. 15 mg/mL GlycoBlue™.
2. 7.5 M Sodium Acetate, pH 5.2.
3. 100% Ethanol.
4. Nuclease-free distilled water.

2.6 Capture of Biotinylated RNA with Magnetic Beads

1. Dynabeads streptavidin.
2. RNA binding buffer: 1 M NaCI, 100 mM Tris-HCI pH 7.0, 10 mM EDTA, 0.2% (v/v) Tween-20.
3. Wash buffer: 0.4 M NaCl, 100 mM Tris-HCl pH 7.0, 10 mM EDTA, 0.2% (v/v) Tween-20.
4. 40 U/μL RiboLock RNase inhibitor.
5. PBS (see Sect. 2.1).

2.7 Elution of Captured RNA

1. Elution buffer: 20 mM Tris pH 7.4, 1 mM EDTA pH 8.0.

2.8 cDNA Synthesis from 5-EU-Labeled RNA and qPCR

1. 5 U/μL Poly (A) polymerase (*see* **Note 3**).
2. Poly (A) tailing buffer: 50 mM Tris-HCl, 250 mM NaCl, 10 mM $MgCl_2$ (pH 8.1).
3. 10 mM ATP.
4. 10 mM dNTP mix.
5. 200 U/μL Reverse transcriptase.
6. 40 U/μL RiboLock RNase inhibitor.
7. Oligo dT adaptor primer:
   - 5′-GCATAGACCTGAATGGCGGTAAGGGTGTGGTA GGCGAGACATTTTTTTTTTTTTTTTTTTT-3′.
8. 10 μM of universal 3′ miRNA reverse primer:
   - 5′-GCATAGACCTGAATGGCGGTA -3′.
9. 10 μM of hsa-*let-7b* forward primer:5′-TGAGGTAGTAG GTTGTGTGGTT-3′.
10. SYBR green master mix.

3 Methods

3.1 5-EU Labeling of Nascent RNA

1. Seed donor cells in 100-mm or 150-mm culture dish and grow until ~50% confluence (*see* **Note 4**).
2. Replace the medium with serum-free conditioned medium.
3. Add 5-EU stock solution (0.2 mM final) into the medium (*see* **Note 5**).
4. Incubate the cells at 37°C for 24 h to allow 5-EU incorporation into nascent RNA (*see* **Note 5**).

3.2 Isolation of Extracellular Vesicles

1. Collect the serum-free conditioned medium into 15 mL tube.
2. Centrifuge at 500 × *g* for 20 min at room temperature to remove cells and debris.
3. Transfer the supernatant to a fresh tube without disturbing the pellet.
4. Add benzonase (1 U/mL final) and incubate for 15 min at room temperature.
5. Centrifuge at 2000 × *g* for 20 min at 25°C.
6. Transfer the supernatant to a fresh tube and centrifuge at 10,000 × *g* for 1 h at 25°C

7. Filter the supernatant through 0.2 μm asymmetric polyether sulfone membrane.
8. Perform tangential flow filtration (TFF) using a 500-kDa cut-off Biomax cassette to concentrate the filtrate. Set the flow rate to 4–5 mL/min using a poly pump, and maintain inlet and operating pressures below 68,947.6 Pa and 24,131.66 Pa, respectively.
9. Wash the retained particles with PBS and further concentrate it to tenfold to 15-fold of the original volume. The volume of final processed sample was typically concentrated tenfold to 15-fold.
10. Centrifuge EV-containing retentate at 200,000 ×*g* for 1 h at 4°C.
11. Resuspend the EVs pellet with PBS and store at −80°C until further use.

3.3 Transfer of Donor Cell-Derived EVs to Recipient Cells and RNA Extraction

1. Seed recipient cells in culture dish and grow until ~70% confluence.
2. Replace the medium with serum-free conditioned medium and incubate for 1 h at 37°C.
3. Add isolated EVs to the medium and incubate for 24 h.
4. Wash the cells twice with PBS.
5. Add 1 mL of Trizol to lyse the cells.
6. Collect the lysate using a cell scraper and transfer it into 1.5 mL tube.
7. Add 200 μL of chloroform and vortex vigorously.
8. Incubate the mixture at room temperature for 3 min.
9. Centrifuge at 12,000 ×*g* for 15 min at 4°C.
10. Transfer the upper aqueous phase containing the RNA to a new tube.
11. Add 500 μL of ice-cold isopropanol to the aqueous phase, invert several times to mix, and incubate for 10 min.
12. Centrifuge at 12,000 ×*g* for 10 min at 4°C and discard the supernatant.
13. Wash the pellet with 1 mL ice-cold 75% ethanol.
14. Invert the tube briefly to mix, then centrifuge for at 7500 ×*g* for 5 min at 4°C. Discard the supernatant.
15. Air-dry the RNA pellet for 10 min in clean-bench. Resuspend in add 30 μL nuclease-free distilled water (*see* **Note 6**).
16. Measure RNA concentration for the click chemistry reaction (*see* **Note 7**).

3.4 Biotinylation of RNA

1. Prior to use, thaw the stock solutions of biotin azide, THPTA, and sodium ascorbate at room temperature.
2. To prepare the click reaction mix, combine 5 µg of total RNA, 25 µL of 5-EU buffer (1× final), 4 µL of $CuSO_4$ (2 mM final), and 2.5 µL of biotin azide (0.5 mM final). Adjust the final volume to 50 µL with DW (*see* **Note 7**).
3. After adding each component, mix the click reaction mixture gently by pipetting.
4. Add 1.25 µL THPTH and immediately mix it by gently pipetting and incubate for 3 min (*see* **Note 8**).
5. After incubation, add 1.5 µL of sodium ascorbate. It should turn dark brown.
6. Incubate the reaction at room temperature for 1 h with gentle shaking (*see* **Note 9**).

3.5 Precipitation of Biotinylated RNA

1. Add 1 µL of glycoblue, 50 µL of 7.5 M of ammonium acetate, and 700 µL of prechilled 100% ethanol to the biotinylated RNA sample. Mix thoroughly by inverting the tube several times.
2. Incubate the mixture overnight at −80°C to precipitate RNA.
3. Centrifuge at 13,000 × *g* for 30 min at 4°C.
4. Discard the supernatant without disturbing the pellet Add 700 µL of 75% ethanol, briefly invert to mix, and centrifuge at 13,000 × *g* for 5 min at 4°C.
5. Repeat Step 4 once more. Air-dry the RNA pellet for 10 min in clean bench, then resuspend in 30 µL of distilled water. Store the RNA at −20°C.

3.6 Capture of Biotinylated RNAs with Magnetic Beads

1. Prepare 20 µL of Dynabeads in 1.5 mL tubes.
2. Centrifuge the Dynabeads at 2000 × *g* for 1 min at 4°C. Carefully aspirate and discard the supernatant.
3. Wash the beads with 500 µL of RNA binding buffer. Place the tube on a magnetic stand for 1 min and discard the supernatant.
4. Repeat Step 3 for two washes.
5. After the final wash, add 500 µL of wash buffer to prevent drying.
6. In a separate tube, prepare reaction mix by combining 75 µL of RNA binding buffer, 1 µL of RNase inhibitor, 30 µL of RNA (from Step 5 of Sect. 3.5), and 44 µL of nuclease-free distilled water. The final volume is 150 µL.
7. Heat the reaction at 70°C for 5 min.

8. Immediately before using the beads, place them (from Step 5) on a magnetic stand and remove the supernatant. Then, combine washed bead with the RNA binding reaction mix.
9. Incubate the reaction at room temperature for 1 h while gently shaking on a thermomixer (*see* **Note 10**).
10. Wash the bead five times using 200 μL of wash buffer. After each wash, remove the buffer using magnetic stand.
11. Finally, wash the beads three times with 200 μL of 1× PBS. Discard the supernatant.

3.7 Elution of Captured RNA

1. To elute RNA, resuspend the beads in 30 μL of elution buffer and incubated at 92°C for 4 min with gentle agitation.
2. Place the tube on a magnetic stand for 1 min to separate the beads.
3. Collect the supernatant containing the 5-EU labeled RNA. Store it at −20°C.

3.8 cDNA Synthesis from 5-EU-Labeled RNA and qPCR

(microRNA cDNA synthesis and qPCR)

1. For Poly (A) tailing, mix 5 μL of purified RNAs with 2 μL of Poly (A) tailing buffer (2× final), 1 μL of Poly(A) polymerase (0.5 U/μL final), and 2 μL of ATP (2 mM final).
2. Incubate the reaction at 37°C for 1 h, followed by 5 min at 70°C. Briefly centrifuge to collect the contents, and place the tube on ice.
3. For cDNA synthesis, mix 10 μL of the poly A tailing reaction (from Step 2) with 1 μL of oligo dT adaptor (6.67 pmol/μL), 1 μL of 10 mM dNTP (0.67 mM final), and nuclease-free DW to a total volume of 15 μL.
4. Incubate the mixture at 65°C for 5 min, then immediately chill at 4°C for 5 min.
5. Add 4 μL of 5× RT buffer (1× final), 0.5 μL of RiboLock RNase inhibitor (1 U/μL), and 0.5 μL reverse transcriptase (5 U/μL) to the mixture.
6. Incubate at 25°C for 10 min, 50°C for 30 min, 85°C for 5 min, and finally at 4°C for 10 min. Store the synthesized cDNA at −20°C until further use.
7. To detect *let-7b* miRNA, mix 2.5 μL of cDNA with 2.5 μL of *let-7b* forward primer (1.25 μM final), 2.5 μL of universal 3′ miRNA reverse primer (1.25 μM final), and 2.5 μL of SYBR green master mix.
8. Perform qPCR and calculate Ct values of each sample (see Fig. 2).

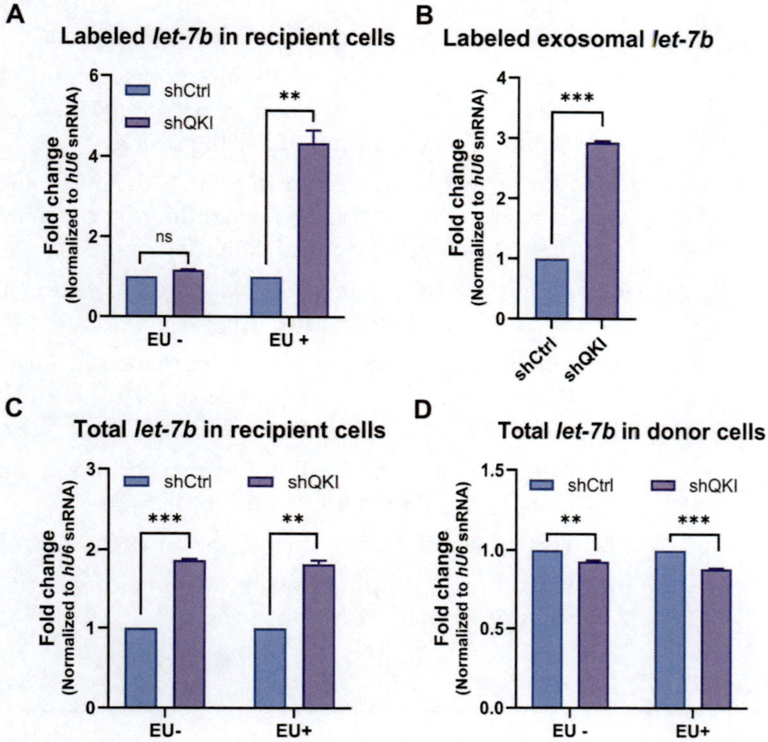

Fig. 2 QKI knockdown increases EV-mediated transfer of *let-7b* from donor to recipient cells. (**A**) Labeled *let-7b* in recipient cells following incubation with EVs isolated from 5-EU-labeled QKI-depleted (shQKI) or control (shCtrl) donor cells. Total RNA from recipient cells was subjected to click-biotinylation, enriched using streptavidin, and analyzed by RT-qPCR. (**B**) Labeled *let-7b* in EVs isolated from 5-EU-labeled donor cells. RNA extracted from EVs was subjected to click-biotinylation, followed by streptavidin enrichment and RT-qPCR analysis. (**C**) Total *let-7b* in recipient cells after EV treatment. Total RNA was extracted from recipient cells and analyzed by RT-qPCR without biotinylation. (**D**) Total *let-7b* in donor cells after EV isolation. Following removal of EVs, total RNA was extracted from donor cells and analyzed by RT-qPCR without biotinylation. Data in (**A–D**) were normalized to h*U6* snRNA and are presented as mean ± SD from $n = 2$. Statistical significance is determined using Student's *t*-test; *ns* not significant; ** $p < 0.01$; *** $p < 0.001$. The figure is adapted from [15], with adjustments to the color palette and panel titles to improve readability.

4 Notes

1. The medium can be replaced with culture conditions suitable for the donor and recipient cell types used in your study.
2. Repeat freeze-thaw cycles as needed. Discard the sample if it turns yellow.
3. EVs contain not only miRNAs but also other types of RNAs. If EV-transferred mRNA analysis is desired, cDNA can be

synthesized from 5-EU-labeled RNA using oligo(dT) primers without poly A tailing, followed by qPCR.

4. Please see the necessary requirements for the growth of the cells of interest. Cells should be seeded to reach 40%–50% confluency at the time of 5-EU pulsing.
5. Optimize EU concentrations (0.1–0.5 mM) and pulse labeling time (30 min to 24 h) depending on the growth medium, cell density, cell type, and other factors.
6. This is a stop point in the experiment. Store the resuspend RNA at −80°C for future use.
7. For the click reaction, it is recommended to use 5 μg of total RNA in a final reaction volume of 50 μL. If sufficient RNA is not available, a minimum of 2–3 μg should be used.
8. Add THPTA and sodium ascorbate at the final last. The $CuSO_4$: THPTH ratio should be 1:5.
9. You may optimize an incubation time from 30 min to 2 h.
10. Wash beads using tenfold the volume of wash buffer relative to bead suspension.

Acknowledgments

J.W.L., C.W.H., and K.-W.M. were supported by the National Research Foundation of Korea (NRF) grant funded by the Korea government (MSIT) (No. RS-2025-16065575).

References

1. Kalluri R, LeBleu VS (2020) The biology, function, and biomedical applications of exosomes. Science 367(6478):eaau6977. https://doi.org/10.1126/science.aau6977
2. Jeppesen DK et al (2019) Reassessment of exosome composition. Cell 177(2):428–445. e418. https://doi.org/10.1016/j.cell.2019.02.029
3. Tsering T et al (2024) Extracellular vesicle-associated DNA: ten years since its discovery in human blood. Cell Death Dis 15(9):668. https://doi.org/10.1038/s41419-024-07003-y
4. Liu T et al (2018) EVmiRNA: a database of miRNA profiling in extracellular vesicles. Nucleic Acids Res 47(D1):D89–D93. https://doi.org/10.1093/nar/gky985
5. Lucero R et al (2020) Glioma-derived miRNA-containing extracellular vesicles induce angiogenesis by reprogramming brain endothelial cells. Cell Rep 30(7):2065–2074. e2064. https://doi.org/10.1016/j.celrep.2020.01.073
6. Dokhanchi M et al (2021) Colorectal cancer cell-derived extracellular vesicles transfer miR-221-3p to promote endothelial cell angiogenesis via targeting suppressor of cytokine signaling 3. Life Sci 285:119937. https://doi.org/10.1016/j.lfs.2021.119937
7. O'Brien K et al (2020) RNA delivery by extracellular vesicles in mammalian cells and its applications. Nat Rev Mol Cell Biol 21(10):585–606. https://doi.org/10.1038/s41580-020-0251-y
8. Mukherjee K et al (2016) HuR-microRNA binding controls extracellular export of miR-122 and augments stress response. EMBO Rep 17(8):1184–1203. https://doi.org/10.15252/embr.201541930
9. Statello L et al (2018) Identification of RNA-binding proteins in exosomes capable of interacting with different types of RNA: RBP-facilitated

transport of RNAs into exosomes. PLOS One 13(4):e0195969. https://doi.org/10.1371/journal.pone.0195969
10. Santangelo L et al (2016) The RNA-binding protein SYNCRIP Is a component of the hepatocyte exosomal machinery controlling MicroRNA sorting. Cell Rep 17(3):799–808. https://doi.org/10.1016/j.celrep.2016.09.031
11. Marocco F et al (2025) Negative regulation of miRNA sorting into EVs is mediated by the capacity of RBP PCBP2 to impair the SYNCRIP-dependent miRNA loading. eLife 14:RP105017. https://doi.org/10.7554/eLife.105017
12. Garcia-Martin R et al (2022) MicroRNA sequence codes for small extracellular vesicle release and cellular retention. Nature 601(7893):446–451. https://doi.org/10.1038/s41586-021-04234-3
13. Singh S et al (2024) Role of extracellular vesicle microRNAs and RNA binding proteins on glioblastoma dynamics and therapeutics development. Extracell. Vesicle 4:100049. https://doi.org/10.1016/j.vesic.2024.100049
14. McKenzie AJ et al (2016) KRAS-MEK signaling controls Ago2 sorting into exosomes. Cell Rep 15(5):978–987. https://doi.org/10.1016/j.celrep.2016.03.085
15. Min KW et al (2024) Mature microRNA-binding protein QKI suppresses extracellular microRNA let-7b release. J Cell Sci 137(21):jcs261575. https://doi.org/10.1242/jcs.261575
16. Bonacquisti EE et al (2024) Fluorogenic RNA-based biomaterials for imaging and tracking the cargo of extracellular vesicles. J Controlled Release 374:349–368. https://doi.org/10.1016/j.jconrel.2024.07.043
17. Morales-Kastresana A et al (2017) Labeling extracellular vesicles for nanoscale flow cytometry. Sci Rep 7(1):1878. https://doi.org/10.1038/s41598-017-01731-2
18. Oliveira GP et al (2020) Detection of extracellular vesicle RNA using molecular beacons. iScience 23(1):100782. https://doi.org/10.1016/j.isci.2019.100782
19. Padilla J-CA et al (2025) Inhibition of acid or neutral sphingomyelinases differentially impacts RNA and protein cargo sorting to extracellular vesicles. iScience 28(5):112440. https://doi.org/10.1016/j.isci.2025.112440
20. Abels ER et al (2019) Glioblastoma-associated microglia reprogramming is mediated by functional transfer of extracellular miR-21. Cell Rep 28(12):3105–3119.e3107. https://doi.org/10.1016/j.celrep.2019.08.036
21. Probert C et al (2019) Communication of prostate cancer cells with bone cells via extracellular vesicle RNA; a potential mechanism of metastasis. Oncogene 38(10):1751–1763. https://doi.org/10.1038/s41388-018-0540-5
22. Castanheira CIGD et al (2025) Extracellular vesicle-derived microRNA crosstalk between equine chondrocytes and synoviocytes—an in vitro approach. Int J Mol Sci 26(7):3353. https://doi.org/10.3390/ijms26073353

Chapter 18

Analysis of Endogenously Expressed Double-Stranded RNAs in Human Cultured Cells

Moonhyeon Jeon and Yoosik Kim

Abstract

Long double-stranded RNAs (dsRNAs) play a critical role in modulating innate immune responses. In particular, endogenously generated dsRNAs are increasingly recognized as a biomarker and effector in immune-related diseases such as cancer, viral infection, and autoimmune disorders. However, accurate detection of endogenous dsRNAs is challenging due to their transient nature and low abundance. This chapter outlines three complementary methods using the dsRNA-specific J2 antibody: J2 dot blot for semiquantitative assessment of dsRNA levels; J2 immunocytochemistry for visualization of dsRNA subcellular localization; and J2 formaldehyde-mediated crosslinking immunoprecipitation coupled with RT-qPCR and/or high-throughput sequencing for quantification and/or identification of dsRNA species. Together, these methods enable sensitive and reproducible analysis of dynamic dsRNA landscapes under pathophysiological conditions, providing a versatile workflow for studying dsRNA signaling in mammalian cell biology and human disease models.

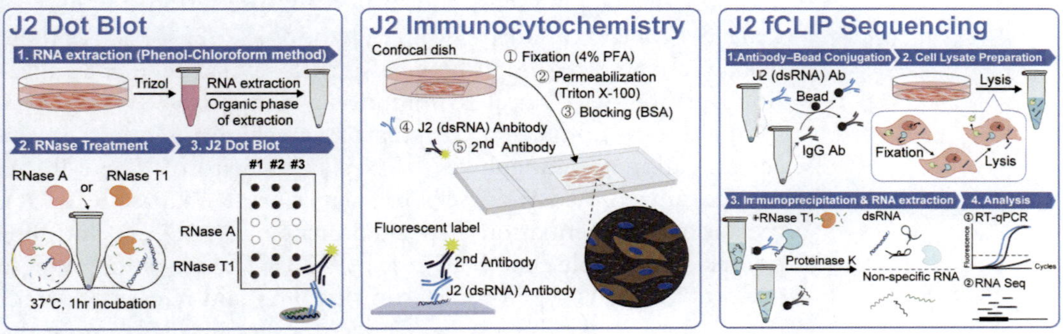

Graphical abstract . Three complementary methods that detect, visualize, and classify endogenous dsRNAs using the J2 antibody, with appropriate RNase controls for enhanced specificity and relative quantitative readouts.

Keywords Double-stranded RNAs, Innate immunity, J2 formaldehyde-mediated crosslinking immunoprecipitation, Immunocytochemistry, Dot blot

Je-Hyun Yoon (ed.), *RNA Labeling*, Methods in Molecular Biology, vol. 3047,
https://doi.org/10.1007/978-1-0716-5352-4_18,

1 Introduction

Living organisms possess a fundamental ability to detect double-stranded RNAs (dsRNAs) to initiate a defense against invaders, such as viruses [1–3]. While widely recognized as a characteristic feature of viral RNA replication, dsRNAs can also be expressed naturally at varying levels in uninfected cells, as these RNAs can be generated from repeat elements of the cellular genome and from mitochondria [4–7]. These intracellular dsRNAs, termed endogenous or cellular dsRNAs, can elicit immune responses as they are mistakenly recognized as foreign due to the structure-based recognition of pattern recognition receptors [8]. Consequently, cells have evolved sophisticated mechanisms to prevent the aberrant activation of dsRNA sensors by self-derived dsRNAs [9, 10].

Endogenous dsRNAs originate from several distinct genomic elements [11, 12]. For example, inverted *Alu* repeats (IR*Alus*) in introns and 3′ untranslated regions (UTRs) can fold back on themselves to form stable intramolecular duplexes [13–16]. Similarly, bidirectional transcription of the mitochondrial genome produces complementary transcripts that pair into mitochondrial dsRNA intermediates, which are normally cleared by the mitochondrial degradosome [17–19]. Under epigenetic stress or following DNA demethylation, reactivated endogenous retroviral elements give rise to retrotransposon RNAs that anneal into long dsRNAs [20, 21]. In addition, overlapping sense and antisense transcription at individual gene loci contributes further duplex hybrids [22–25]. Each of these dsRNA classes differs in its biogenesis pathway, subcellular localization, and ability to stimulate immune signaling [26–28].

Failure of homeostatic regulatory mechanisms results in the accumulation of endogenous dsRNAs in the cytosol. These RNAs are recognized by dsRNA sensors such as protein kinase R (PKR), melanoma differentiation-associated protein 5 (MDA5), and retinoic acid–inducible gene I (RIG-I) [29–33]. The activation of these pathways triggers type I interferon signaling and release of proinflammatory cytokines [33, 34]. Indeed, several studies suggest the misregulation of dsRNAs as a potential driver for the development of autoinflammatory and degenerative disorders, including osteoarthritis, Aicardi–Goutières syndrome, and Singleton–Merten syndrome [35, 36]. In the nervous system, excessive mitochondrial dsRNAs and IR*Alu*s have been implicated in the progression of Huntington's disease, amyotrophic lateral sclerosis, and frontotemporal dementia [37, 38]. In contrast, in cancer, DNA-demethylating agents can reactivate the transcription of endogenous retroviruses, creating a viral-mimicry state that enhances antitumor immunity [20, 21, 39].

Despite their potential clinical values as pathophysiological markers of immune-related diseases, endogenous dsRNAs are difficult to detect due to their low abundance and transient expression, requiring highly sensitive and specific detection methods [1, 40]. Here, we describe three complementary assays based on the dsRNA-specific J2 antibody for the detection and analysis of dsRNAs in cultured human cells. First, the J2 dot blot allows for a quick, semiquantitative analysis of overall dsRNA expression in cell extracts (Fig. 1). However, the J2 dot blot fails to provide spatial information on dsRNA expression. This can be resolved through immunocytochemistry (J2 ICC), which enables the visualization and analysis of colocalization between dsRNAs and specific cellular compartments, such as stress granules and organelles (Fig. 2). Moreover, the fluorescence signal from the J2 ICC can be quantified to obtain relative expression

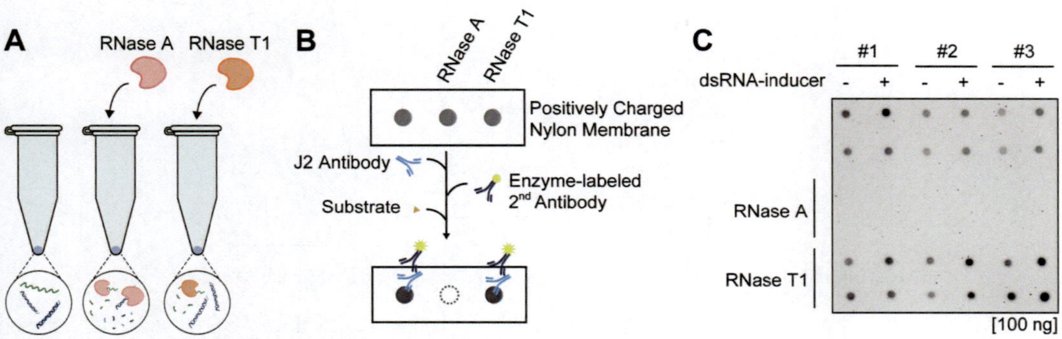

Fig. 1 J2 dot blot assay. (**A**) Experimental design with appropriate RNase controls. Total RNA is divided into three conditions: untreated, RNase A–treated, and RNase T1-treated. (**B**) Workflow for the J2 dot blot assay. (**C**) A dsRNA-inducing treatment increases the J2 signal relative to the control. Each sample is spotted in technical duplicate, and three biologically independent RNA preparations are analyzed ($n = 3$). Decreased signal intensity after RNase T1 indicates contribution by ssRNAs. The complete loss of signal after RNase A treatment reflects its activity on both ssRNAs and dsRNAs

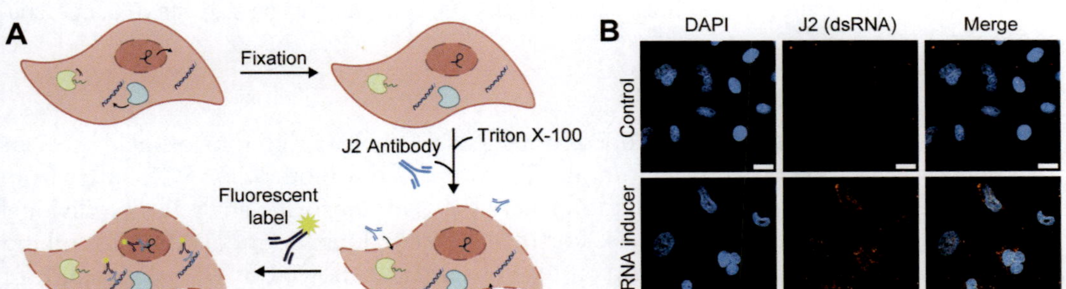

Fig. 2 J2 immunocytochemistry. (**A**) Schematic overview of the workflow. Cells are fixed with paraformaldehyde, permeabilized with Triton X-100, and blocked to reduce nonspecific binding. They are then incubated with the J2 antibody, followed by incubation with a fluorophore-conjugated secondary antibody. (**B**) Representative confocal images of control versus dsRNA-induced cells. Nuclei are counterstained with DAPI, and images are acquired by confocal microscopy with a 63× oil-immersion objective. Treatment with a dsRNA inducer increases both cytoplasmic and nuclear J2 signals. Scale bar, 20 μm

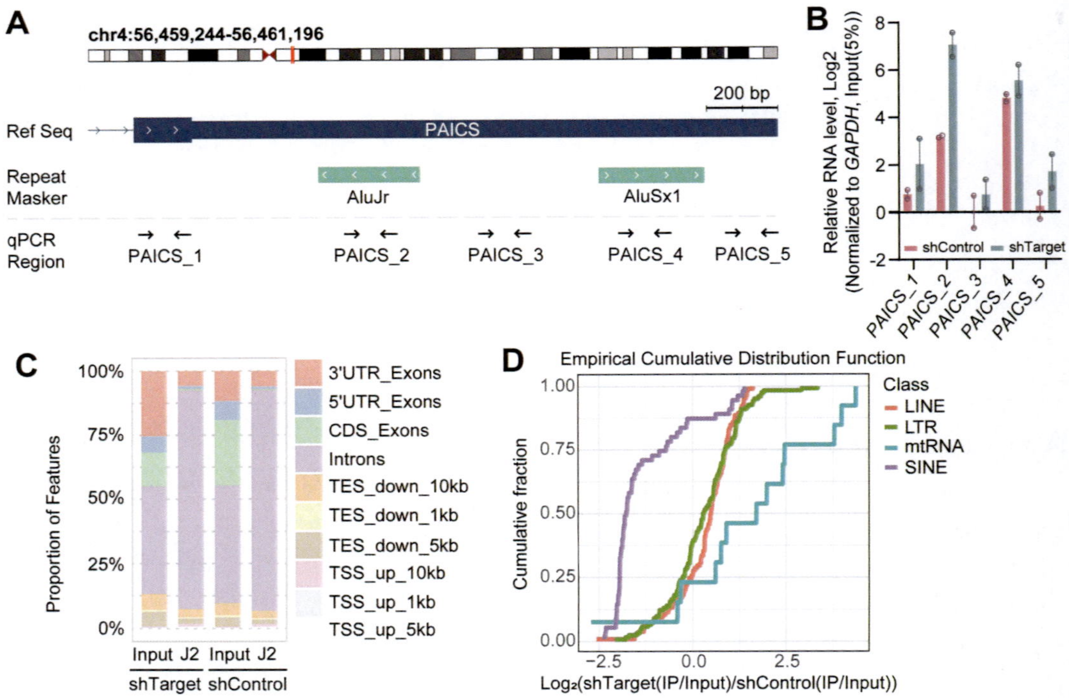

Fig. 3 Analysis of dsRNA species captured by J2. (**A**) Primer design across the *PAICS* locus, which is used as a representative IR*Alus*-containing mRNA to test the quality of RNAs capture by J2 antibody. Primer sets #2 and #4 target regions predicted to form a duplex between inverted *Alu* elements and are expected to be enriched in J2 immunoprecipitates. Primer sets #1, #3, and #5 target flanking nonduplex regions and serve as negative controls. (**B**) RT-qPCR of J2 IP and matched input. Primer sets #2 and #4 show increased signal in J2 IP, whereas sets #1, #3, and #5 do not. Data are shown as mean ± SEM. From technical duplicates per biological replicate ($n = 2$). (**C**) J2 fCLIP RNA-seq profiling of genomic compartments. Bars depict the proportion of enriched reads assigned to 5′ UTR, CDS exons, 3′ UTR, introns, and TSS/TES flanks. (**D**) Empirical cumulative distribution function (ECDF) plots showing the distribution of Log_2 fold changes in J2-IP enrichment after target-gene knockdown relative to control. Data shown in panels (**C**) and (**D**) are reanalyzes of publicly available J2 fCLIP RNA-seq data (Gene Expression Omnibus, GSE297870). TSS, transcription start site; TES, transcription end site

changes of endogenous dsRNAs across different cellular contexts or stress conditions. Yet, both J2 dot blot and J2 ICC suffer from a critical limitation where they fail to provide analysis on individual dsRNA species. For this, J2 formaldehyde-mediated crosslinking immunoprecipitation (J2 fCLIP) followed by RT-qPCR and/or high-throughput sequencing can be employed to quantitatively analyze individual dsRNA species (Fig. 3).

By integrating selective RNase controls, these assays constitute a cohesive pipeline for analyzing endogenous dsRNAs in mammalian cells. For example, RNase T1 trimming confirms duplex integrity by eliminating the single-stranded region of the RNA. In addition, the J2 antibody directly binds dsRNAs with high affinity, enabling its detection across various workflows, such as dot blot, immunocytochemistry, immunoprecipitation, and sequencing. Finally, J2 fCLIP enriches endogenous dsRNA duplexes for sequence-resolved profiling, allowing discrimination among dsRNA classes, such as inverted *Alu* repeats (IR*Alus*), mitochondrial dsRNAs, and endogenous retrovirus (ERV)-derived RNA duplexes. This versatile platform can be used to profile dynamic dsRNA changes during pathological conditions, allowing us to screen modulators of dsRNA regulators and their downstream effectors, as well as identify therapeutic targets in dsRNA-associated diseases.

2 Materials

1. A549 (ATCC, Cat#CCL-185)
2. RPMI 1640 Medium (Welgene, Cat#LM011-01)
3. Fetal Bovine Serum, qualified, United States (Gibco, Cat#26140079)

2.1 J2 Dot Blot

1. TRIzol™ Reagent (Invitrogen, Cat#15596026)
2. 3M Sodium Acetate pH 5.5 (Invitrogen, Cat#AM9740)
3. Glycoblue Coprecipitant (Invitrogen, Cat#AM9516)
4. Recombinant DNase I (RNase-free) (Takara, Cat#2270B)
5. Recombinant RNase inhibitor (Takara, Cat#2313B)
6. Acid-Phenol:Chloroform, pH 4.5 (Invitrogen, Cat#AM9722)
7. RNase T1 (Worthington Biochemical, Cat#LS01490)
8. RNase A (Worthington Biochemical, Cat#LS003431)
9. Amersham Hybond-N+ nylon membrane (Cytiva, Cat#RPN1210B)
10. Skim milk powder (Millipore, Cat#115363)
11. J2 (anti-dsRNA) mouse monoclonal antibody (English and Scientific Consulting Kft, Cat#10010200)
12. Goat Anti-Mouse IgG HRP (Jackson Immunoresearch, Cat#115-035-146)
13. Clarity Western ECL Substrate (Bio-Rad, Cat#1705061)
14. 0.1% TBST: 24.7 mM Tris base, 137 mM NaCl, 2.7 mM KCl, 0.1% (v/v) Tween-20

2.2 J2 Immunocytochemistry

1. Gelatin (MP Biomedicals, Cat#901771) (*see* **Note 1**)
2. Paraformaldehyde (Sigma-Aldrich, Cat#158127-100 G) (*see* **Note 2**)

3. Triton X-100, Molecular Biology Grade (Promega, Cat#H5141)
4. Phosphate-Buffered Saline (PBS), pH 7.4 (Takara, Cat#T9181)
5. 10% Tween-20 (Bio-Rad, Cat#1610781)
6. Sodium Dodecyl Sulfate (Bioworld, Cat#420203101)
7. Bovine Serum Albumin (RMBIO, Cat#BSA-BSH)
8. J2 (anti-dsRNA) mouse monoclonal antibody (English and Scientific Consulting Kft, Cat#10010200)
9. Donkey mouse IgG (H+L) secondary antibody, Alex Fluor 555, 1000× (Invitrogen, Cat#A-31570)
10. 4′,6-Diamidino-2-Phenylindole, Dihydrochloride (DAPI) (Invitrogen, Cat# D1306)
11. SlowFade Diamond Antifade Mountant (Invitrogen, Cat#S36972)
12. 0.1% PBST: 0.1% (v/v) Tween-20 in PBS

2.3 J2 Formaldehyde-Mediated Crosslinking Immunoprecipitation

1. Pierce™ Protein A Plus Agarose (Thermo Fisher Scientific, Cat#22810)
2. J2 (anti-dsRNA) mouse monoclonal antibody (English and Scientific Consulting Kft, Cat#10010200)
3. Protease inhibitor cocktail (Thermo Fisher Scientific, Cat#87785)
4. Sodium deoxycholate (Sigma-Aldrich, Cat#D6750)
5. Dithiothreitol (DTT) (VWR Life Science, Cat#VWRV0281-25G)
6. dNTP mixture (2.5 mM) (Takara, Cat#4030)
7. Glycine (Bio-basic, Cat#GB0235)
8. NP-40 (Biosolution, Cat#BN015)
9. 20% SDS (Biosolution, Cat#BS003)
10. 1 M Tris-HCl pH 7.0 (Invitrogen, Cat#AM9850G)
11. 1 M Tris-HCl pH 8.0 (Invitrogen, Cat#AM9855G)
12. 0.5 M EDTA (Invitrogen, Cat#AM9260G)
13. Proteinase K, recombinant, PCR Grade (Roche, Cat#3115879001)
14. SuperScript IV Reverse Transcriptase (Invitrogen, Cat#18090200)
15. Random hexamer primer (Thermo Fisher Scientific, Cat#SO142)

16. SensiFAST SYBR Lo-ROX Kit (Bioline, Cat#BIO-94020)
17. RNase T1 (Worthington Biochemical, Cat#LS01490)
18. 5 M NaCl, RNase-free (Invitrogen, Cat#AM9760G)
19. Urea (Sigma-Aldrich, Cat#U5378-500G)
20. 0.1% PBST: 0.1% (v/v) Tween-20 in PBS
21. fCLIP lysis buffer: 20 mM Tris-HCl (pH 7.5), 15 mM NaCl, 10 mM EDTA, 0.5% (v/v) NP-40, 0.1% (v/v) Triton X-100, 0.1% (v/v) SDS, 0.1% (w/v) sodium deoxycholate
22. fCLIP wash buffer: 20 mM Tris-HCl (pH 7.5), 150 mM NaCl, 10 mM EDTA, 0.1% (v/v) NP-40, 0.1% (v/v) Triton X-100, 0.1% (v/v) SDS, 0.1% (w/v) sodium deoxycholate
23. fCLIP elution buffer: 200 mM Tris-HCl (pH 7.5), 100 mM NaCl, 20 mM EDTA, 2% (v/v) SDS, 7 M urea
24. PAICS qPCR primer

	Forward (5′-3′)	Reverse (5′-3′)
PAICS-1	ACCAAGATATTTCAGCCAGCCT	GCTGGGAGTCAGTAATGCCA
PAICS-2	CCTCTCGACTAACAGGGACTATAA	AGGAGAATTACTAGGGTTCAGGAG
PAICS-3	TCCTGATCTAGACAGCCCCC	TGCGGTTTTCTACAGAGAGGG
PAICS-4	TACGATCACAGCACTTCGGG	GTCATCACGGCTGGCTACTT
PAICS-5	GTTCTCTGAAGGCCCACGTT	GGTTTCATGGACAGCGAACA

3 Methods

3.1 J2 Dot Blot

3.1.1 RNA Extraction (see Note 3)

1. After trypsinization, seed cells into a 6-well plate with culture medium (*see* **Note 4**).
2. Incubate cells in a humid 37°C incubator with 5% CO_2.
3. Aspirate the culture medium.
4. Rinse the cells once with PBS and remove the PBS completely by suction.
5. Add 1 mL/well of TRIzol directly onto cells grown in a 6-well plate.

6. Suspend the solution by pipetting up and down until all cells are dissolved in TRIzol and transfer the solution into a 1.7 mL microcentrifuge tube.
7. Add 200 μL of chloroform and vortex the solution thoroughly for 30 s.
8. Centrifuge the sample at 13,523 × *g* for 15 min at 4°C.
9. Carefully transfer the 500 μL supernatant to a fresh 1.7 mL microcentrifuge tube (*see* **Note 5**).
10. Add 550 μL of ice-cold isopropanol (110% of the supernatant volume), 50 μL of 3 M sodium acetate (pH 5.5; 10% of the supernatant volume), and 0.5 μL of GlycoBlue to the supernatant (*see* **Note 6**).
11. Vortex briefly for 5 s, then incubate at –20°C for 18 h.
12. Centrifuge the sample at 21,130 × *g* for at least 1 h at 4°C.
13. Remove the supernatant and wash the pellet with 1 mL of 75% ethanol at 13,523 × *g* for 5 min at 4°C.
14. Remove the supernatant and repeat the 75% ethanol wash. After the final wash, perform a brief spin to collect residual liquid at the bottom of the tube, and then carefully aspirate any remaining ethanol.
15. Air-dry the pellet for 15 min at room temperature (*see* **Note 7**).
16. Add 42 μL of RNase-free water to the sample and dissolve the pellet thoroughly by pipetting.
17. Prepare the DNase I mixture by combining 5 μL of 10× DNase I buffer, 1 μL of RNase inhibitor, and 2 μL of DNase I in a fresh 1.7 mL microcentrifuge tube.
18. Add the entire 8 μL mixture to the sample.
19. Mix the solution by tapping and spin down the sample.
20. Incubate the sample in a 37°C heat block for 30 min.
21. Add 150 μL of RNase-free water and 200 μL of acid-phenol:chloroform.
22. Vortex vigorously for 30 s.
23. Centrifuge the sample at 13,523 × *g* for 10 min at room temperature.
24. Carefully transfer 200 μL of supernatant to a fresh 1.7 mL microcentrifuge tube.
25. Add 1 mL of 100% ethanol, 20 μL of sodium acetate (pH 5.5; 10% of the supernatant volume), and 0.5 μL of GlycoBlue to the supernatant.
26. Vortex briefly for 5 s and incubate the sample at −80°C for 18 h.

27. Centrifuge the sample at 21,130×*g* for at least 1 h at 4°C.
28. Remove the supernatant and wash the RNA pellet with 75% ethanol at 13,523×*g* for 5 min at 4°C.
29. Remove the supernatant and repeat the 75% ethanol wash. After the final wash, perform a brief spin to collect residual liquid at the bottom of the tube, then carefully aspirate any remaining ethanol.
30. Air-dry the pellet for 15 min at room temperature.
31. Dissolve the RNA pellet in 20 μL of fresh RNase-free water (*see* **Note 8**).
32. Measure RNA concentration and purity using a NanoDrop spectrophotometer. Record the A260/A280 ratio (RNA≈2.0) and the A260/A230 ratio (2.0–2.2) as indicators of purity.

3.1.2 J2 Dot Blot

1. Dilute equal amounts of total RNA in RNase-free water to achieve 100–250 ng per 0.5–1 μL (*see* **Note 9**).
2. Divide each sample into three aliquots.
3. Add RNase T1 to the first aliquot, RNase A to the second, and leave the third untreated.
4. Incubate all samples at 37°C for 1 h.
5. Spot 0.5 μL of each treated and untreated RNA onto a positively charged nylon membrane, leaving approximately 1 cm of space between each dot (*see* **Note 10**).
6. Air-dry the membrane for 1 h at room temperature (*see* **Note 11**).
7. Immerse the membrane in TBST and keep it wet from this point onward.
8. Block the membrane in 5% skim milk in TBST for 1 h at room temperature with gentle agitation.
9. Incubate the membrane overnight at 4°C with J2 primary antibody diluted at a 1:500 ratio in 1% skim milk (*see* **Note 12**).
10. Wash the membrane three times with TBST for 10 min each at room temperature.
11. Incubate with HRP-conjugated anti-mouse secondary antibody diluted at a 1:3000 ratio in 1% skim milk for 1 h at room temperature.
12. Wash the membrane four times with TBST for 10 min each at room temperature.
13. Apply enhanced chemiluminescence reagents according to the manufacturer's instructions.

14. Capture the signal with a chemiluminescence imaging system (*see* **Note 13**).
15. Measure signal intensities in ImageJ and average at least three technical replicates per condition.

3.2 J2 Immunocytochemistry (see Note 14)

1. Add 1 mL of 0.1% (w/v) gelatin in deionized water to each confocal dish.
2. Incubate at 37°C with 5% CO_2 overnight.
3. Aspirate the gelatin and rinse twice with pre-warmed culture medium (37°C).
4. Seed 1.0×10^5 A549 cells in 1 mL of culture medium per dish to reach 70%–80% confluence.
5. Incubate at 37°C with 5% CO_2 for 24 h.
6. Rinse cells twice with prewarmed PBS (37°C).
7. Add 1 mL of freshly prepared 4% (w/v) paraformaldehyde in PBS.
8. Incubate for 15 min at room temperature.
9. Aspirate fixative and rinse cells twice with PBS.
10. Wash cells three times with PBS for 5 min each.
11. Add 1 mL of 0.1% (v/v) Triton X-100 in PBS.
12. Incubate for 10 min at room temperature.
13. Wash cells twice with PBST for 5 min each.
14. Add 1 mL of 1% (w/v) bovine serum albumin in PBST (filter through a 0.22 μm membrane).
15. Incubate for 1 h at room temperature.
16. Remove the blocking solution.
17. Add anti-dsRNA J2 antibody diluted 1:100 in 1% BSA.
18. Incubate overnight at 4°C in a chamber to prevent drying.
19. Rinse cells twice with PBST.
20. Wash cells three times with PBST for 5 min each at room temperature.
21. Add fluorophore-conjugated secondary antibody and 450 nM DAPI in 1% BSA.
22. Incubate for 1 h at room temperature in the dark.
23. Rinse cells twice with PBST.
24. Wash cells three times with PBST for 5 min each at room temperature.
25. Carefully detach the coverslip from the confocal dish.
26. Place 10–20 μL of antifade mounting medium on a microscope slide, then invert the coverslip onto the mountant.

27. Seal the edges with clear nail polish and air-dry for 5 min at room temperature.
28. Store slides in a light-protected container at 4°C overnight prior to imaging.
29. Acquire images using a confocal microscope with a 63× oil-immersion objective.
30. Analyze dsRNA signal intensity in ImageJ.

3.3 J2 Formaldehyde-Mediated Crosslinking Immunoprecipitation

3.3.1 Antibody–Bead Conjugation (Performed on Ice)

1. Pipette 20 μL of Protein A agarose beads into a 1.7 mL microcentrifuge tube (*see* **Note 15**).
2. Add 400 μL of fCLIP lysis buffer and invert the tube four times.
3. Centrifuge at 1000 × *g* for 1 min at 4°C.
4. Remove the supernatant and repeat the wash two more times.
5. Add 300 μL of fCLIP lysis buffer, adjust NaCl to 150 mM by adding 12 μL of 5 M NaCl, then add J2 antibody to a final concentration of approximately 30 μg/mL (*see* **Note 16**).
6. Rotate at 4°C for 3 h using a gentle rotator.
7. Pellet beads at 1000 × *g* for 1 min at 4°C and remove supernatant.
8. Add 400 μL of fCLIP wash buffer and gently invert the tube four times (*see* **Note 17**).
9. Centrifuge at 1000 × *g* for 1 min at 4°C.
10. Remove the supernatant and repeat the wash once more. After the final wash, perform a brief spin to collect any residual buffer from the tube walls, then carefully remove all remaining liquid.

3.3.2 Cell Lysate Preparation (Performed on Ice)

1. Harvest cells by scraping the cell culture dish (*see* **Note 18**).
2. Centrifuge the cells at 380 × *g* for 3 min at room temperature and aspirate the medium.
3. Resuspend cells in 1 mL ice-cold PBS and transfer to a microcentrifuge tube.
4. Pellet cells at 10,000 × *g* for 30 s at 4°C and remove supernatant.
5. Add 750 μL of 0.1% paraformaldehyde in PBS and gently resuspend the cell pellet by pipetting.
6. Incubate for 10 min at room temperature on a rotator.
7. Add 250 μL of 1 M glycine to quench (final concentration of 0.25 M).
8. Incubate for 10 min at room temperature on a rotator.

9. Centrifuge the sample at 10,000 × *g* for 1 min at 4°C and remove supernatant.
10. Wash the pellet with 1 mL PBS, centrifuge at 10,000 × *g* for 1 min at 4°C and remove supernatant.
11. Briefly spin down to collect residual liquid on the tube walls and carefully remove any remaining buffer.
12. Resuspend the pellet in 400 μL of fCLIP lysis buffer supplemented with 1 × protease inhibitor cocktail and 1 mM DTT.
13. Incubate on ice for at least 10 min.
14. Sonicate on ice for two rounds of 10 cycles, 30 s on and 30 s off per cycle (*see* **Note 19**).
15. Add 12 μL of 5 M NaCl to reach 150 mM, vortex briefly.
16. Centrifuge at 21,130 × *g* for 15 min at 4°C.
17. Transfer the supernatant to a fresh 1.7 mL microcentrifuge tube.
18. Add RNase T1 to the lysate at a final concentration of 50 U/mL (12 nM).

3.3.3 Immunoprecipitation

1. Combine 300 μL of clarified lysate with the antibody-conjugated beads and reserve the remaining ~100 μL as the input sample.
2. Rotate at 4°C for 3 h using a gentle rotator.
3. Pellet beads at 1000 × *g* for 1 min at 4°C and remove supernatant.
4. Wash the beads with 400 μL of fCLIP wash buffer by gently adding the buffer along the tube wall to avoid disturbing the bead-antibody complexes, then mix by gentle inversion.
5. Centrifuge at 1000 × *g* for 1 min at 4°C and remove the supernatant.
6. Repeat this wash step three additional times for a total four washes. After the final wash, ensure complete removal of all wash buffer to minimize background signal.
7. Add 300 μL of freshly prepared fCLIP elution buffer at room temperature (*see* **Note 20**).
8. Incubate at 25°C for 3 h on a thermomixer at 900 rpm.
9. Pellet beads at 1000 × *g* for 1 min, then transfer the elute to a fresh siliconized tube.
10. Add 300 μL of proteinase K to a final concentration of 2 mg/mL.
11. Incubate overnight at 65°C on a thermomixer at 300 rpm (*see* **Note 21**).
12. Add an equal volume of acid-phenol:chloroform.
13. Vortex vigorously for 30 s.

14. Incubate at 37°C for 1 h in a heat block.
15. Vortex vigorously for 30 s.
16. Centrifuge at 13,523 × *g* for 10 min at room temperature.
17. Transfer the aqueous phase to a fresh siliconized tube.
18. Add 550 μL of ice-cold isopropanol (110% of the supernatant volume), 50 μL of 3 M sodium acetate (pH 5.5; 10% of the supernatant volume), and 0.5 μL of GlycoBlue to the supernatant
19. Incubate at −20°C for at least 16 h.
20. Centrifuge the sample at 21,130 × *g* for at least 2 h at 4°C.
21. Remove the supernatant and wash the pellet twice with 75% ethanol at 13,523 × g for 5 min at 4°C. After the final wash, perform a brief spin to collect residual liquid at the bottom of the tube, then carefully aspirate any remaining ethanol.
22. Air-dry the pellet for 15 min at room temperature
23. Add 42 μL of RNase-free water to the sample and dissolve the pellet thoroughly by pipetting.
24. Prepare the DNase I mix by combining 5 μL of 10 × DNase I buffer, 1 μL of RNase inhibitor, and 2 μL of DNase I.
25. Add 8 μL of this mix to the sample.
26. Mix the solution by tapping and spin down the sample.
27. Incubate the sample in a 37°C heat block for 1 h.
28. Add 150 μL of RNase-free water and 200 μL of acid-phenol:chloroform.
29. Vortex vigorously for 30 s.
30. Centrifuge the sample at 13,523 × *g* for 10 min at room temperature.
31. Carefully transfer 200 μL of supernatant to a fresh 1.7 mL microcentrifuge tube.
32. Add 1 mL of 100% ethanol, 20 μL of sodium acetate (pH 5.5; 10% of the supernatant volume), and 0.5 μL of GlycoBlue to the supernatant.
33. Vortex for 5 s and incubate the sample at −80°C for 18 h.
34. Centrifuge the sample at 21,130 × *g* for at least 2 h at 4°C.
35. Remove the supernatant and wash the RNA pellet with 75% ethanol at 13,523 × *g* for 5 min at 4°C.
36. Remove the supernatant and repeat the wash step.
37. Air-dry for 15 min at room temperature.
38. Resuspend the RNA pellet in an appropriate volume of RNase-free water.

39. Measure RNA concentration and purity using a NanoDrop spectrophotometer

3.3.4 RT-qPCR

1. Prepare the RNA–primer mix in a nuclease-free tube:

Reagent	Amount
Total RNA	10 pg–5 μg
100 μM Random hexamers	0.5 μL
2.5 mM dNTP	4 μL
Nuclease-free water	To 13 μL

2. Mix and briefly spin down the sample.
3. Heat the mixture at 65°C for 5 min, then place on ice for at least 1 min.
4. In a separate tube, assemble the RT reaction mix for each sample:

Reagent	Amount
5× SSIV buffer	4 μL
100 mM DTT	1 μL
Recombinant RNase inhibitor (40 U/μL)	1 μL
SuperScript IV Reverse Transcriptase (200 U/μL)	1 μL

5. Add 7 μL of RT reaction mix to the 13 μL RNA–primer mix (final 20 μL).
6. Mix gently and centrifuge briefly.

Steps	Temperature	Time
Primer extension	23°C	10 min
Reverse transcription	50–55°C	10 min
Inactivation	80°C	10 min
Hold	4°C	Until use

7. Prepare the qPCR master mix as follows (primers should be designed to target sites known to form dsRNA to validate efficient immunoprecipitation; see Fig. 3):

Reagent	Amount
2× SYBR reagent	7.5 μL
Nuclease-free water	4.5 μL
3 μM qPCR primer mix (3 μM each of forward and reverse)	2 μL

8. Dispense 14 μL of the master mix into each well of a 96-well optical plate.
9. Add 1 μL of cDNA template to the corresponding wells
10. Seal the plate with an optical adhesive film and centrifuge at 100 × *g* for 30 s.
11. Place the plate in the real-time PCR instrument.
12. Program the instrument:

Step	Temperature	Time	Cycles
Initial denaturation	95°C	3 min	1
PCR			40
Denaturation	95°C	5 s	
Annealing/extension	60°C	10 s	
	95°C	15 s	1
Melt curve	60°C	1 min	1
	95°C	1 s	1

13. Export Ct values for each *PAICS* primer pair (#1–#5).
14. Normalize each *PAICS* Ct to *GAPDH* (ΔCt = Ct(*PAICS*) – Ct(*GAPDH*)).
15. For each primer pair, calculate enrichment relative to the corresponding input ($\Delta\Delta$Ct = ΔCt(IP) – ΔCt(Input)).
16. Compute fold enrichment as 2^(–$\Delta\Delta$Ct). The IR*Alus*-containing regions (#2 and #4) are expected to show strong enrichment in the J2 IP sample, whereas the non-IR*Alus* control regions (#1, #3, and #5) should not. Confirmation of this pattern by RT-qPCR validates the success of J2 IP, after which samples can be submitted for high-throughput sequencing.

3.3.5 Analysis of High-Throughput Sequencing Data

1. Index the human reference genome.
2. Align paired-end FASTQ files to the reference using a splice-aware aligner (e.g., STAR).
3. Sort and index the resulting alignments by genomic coordinates.
4. Use a transposable element–aware counting tool (e.g., TEtranscripts) in conjunction with a comprehensive gene annotation (e.g., GENCODE/Ensembl GTF) and repeat element annotations, such as RepeatMasker GTF, to generate combined counts for both genes and transposable elements.
5. Compile a raw count matrix encompassing all features across IP and input samples.
6. Filter out low-abundance features to reduce noise.

7. Apply an appropriate normalization strategy to correct for library size and composition differences.
8. Model enrichment between immunoprecipitation and input samples by computing $\log_2$ fold changes with associated statistics using a differential analysis framework.
9. Identify enriched features using predefined thresholds (e.g., ≥ 2-fold change and an FDR-adjusted p value below a threshold), then proceed to visualization and interpretation.

4 Notes

1. Prepare 0.1% (w/v) gelatin in triple distilled water, autoclave before first use, and store at 4°C. Use within 1 month and discard if turbidity or precipitate is observed.
2. Shield 4% (w/v) paraformaldehyde from light by wrapping the tube in aluminum foil. Store at 4°C and use within 1 month.
3. It is essential to remove all RNases and fine dust before starting the experiment by wiping the bench with 10% bleach (sodium hypochlorite) followed by 70% ethanol.
4. Adjust the initial seeding density to reach 80%–90% confluency at the time of cell harvest to avoid over-confluence–induced stress while yielding a sufficient amount of RNA.
5. Avoid touching the middle layer to prevent contamination that may compromise the RNA quality. It is better to use a 200 μL rather than a 1000 μL tip to carefully and thoroughly remove the aqueous layer.
6. Store isopropanol at 4°C until use.
7. Before air-drying, lightly spray 70% ethanol on the bench surface and in the surrounding air to settle dust particles. While air-drying, protect the RNA pellet from fine dust and other contaminants by covering the microtube with plastic wrap. If the pellet is not completely dried, extend the drying time.
8. For downstream applications, such as dot blot analysis, avoid using an excessive volume of RNase-free water, as diluted RNA may reduce detection sensitivity. To increase RNA concentration, reduce the resuspension volume.
9. Normalize RNA concentration across samples, then load an identical volume per spot (0.5–1 μL). Avoid volumes > 1 μL, which may introduce edge effects. Optimize the mass per spot in advance with a small dilution series (e.g., 100, 200 ng) to define a linear detection window for quantitative

comparison. Choose a loading amount that is above the limit of detection yet below signal saturation and keep the same total mass across samples for valid comparison.

10. Avoid placing dots too close to the edges of the membrane, as dots near the edges may be excessively washed away during subsequent washing steps.
11. Protect the membrane from dust without direct contact.
12. Incubate in a sealed container on a rocker to prevent drying.
13. To verify equal RNA loading across samples, briefly stain the membrane with Methylene Blue after dot blot. This allows visual confirmation that comparable amounts of total RNA were spotted on the membrane, ensuring that signal differences are not due to unequal loading.
14. Avoid letting the cells dry out at any stage. Drying can disrupt cell morphology and damage antigen epitopes, resulting in high background and loss of signal.
15. Prepare Protein A agarose beads not only for the specific antibody but also for an isotype-matched control IgG. Including a control IgG IP allows the distinction between specific and nonspecific interactions. Before distributing the beads into tubes, vortex the bead suspension briefly to ensure that the beads are evenly mixed and that an equal amount of beads is added to each sample.
16. Ensure that the final amount of antibody used for the control IgG IP is matched to that used for the specific antibody IP for a valid comparison.
17. To minimize the antibody detachment from the beads during washing, add the wash buffer slowly along the inner wall of the tube and allow it to flow down gently.
18. Sequencing library preparation requires large initial amounts; for A549 lung adenocarcinoma cells, use three 150 mm dishes per sample.
19. Use a slurry of water and ice to keep samples cold and prevent heat buildup. Maintain the bath between 0°C and 4°C, but do not pack it with ice, as excess ice can dampen cavitation and reduce lysis efficiency.
20. Always prepare fresh elution buffer before use, as prolonged storage can reduce its effectiveness. During elution, keep both the samples and the elution buffer at room temperature rather than on ice.
21. Use a heated lid or sealed tubes to prevent evaporation.

Acknowledgments

This study was supported by the Bio&Medical Technology Development Program of the National Research Foundation (NRF) funded by the Korean government's Ministry of Science and ICT (Grant number: RS-2025-02216696).

References

1. Chen YG, Hur S (2022) Cellular origins of dsRNA, their recognition and consequences. Nat Rev Mol Cell Biol 23(4):286–301
2. Jensen S, Thomsen AR (2012) Sensing of RNA viruses: a review of innate immune receptors involved in recognizing RNA virus invasion. J Virol 86(6):2900–2910
3. Alexopoulou L et al. (2001) Recognition of double-stranded RNA and activation of NF-κB by toll-like receptor 3. Nature 413(6857):732–738
4. Sadeq S et al. (2021) Endogenous double-stranded RNA. Non-coding RNA 7(1):15
5. Ahmad S et al. (2018) Breaching self-tolerance to Alu duplex RNA underlies MDA5-mediated inflammation. Cell 172(4):797–810.e13
6. Chen LL, DeCerbo JN, Carmichael GG (2008) Alu element-mediated gene silencing. EMBO J 27(12):1694–1705
7. Lee K et al. (2024) Inverted Alu repeats: friends or foes in the human transcriptome. Exp Mol Med 56(6):1250–1262
8. Liddicoat BJ et al. (2015) RNA editing by ADAR1 prevents MDA5 sensing of endogenous dsRNA as nonself. Science 349(6252):1115–1120
9. Pestal K et al. (2015) Isoforms of RNA-editing enzyme ADAR1 independently control nucleic acid sensor MDA5-driven autoimmunity and multi-organ development. Immunity 43(5):933–944
10. Jiao H et al. (2022) ADAR1 averts fatal type I interferon induction by ZBP1. Nature 607(7920):776–783
11. Cottrell KA, Andrews RJ, Bass BL (2024) The competitive landscape of the dsRNA world. Mol Cell 84(1):107–119
12. Hur S (2019) Double-stranded RNA sensors and modulators in innate immunity. Annu Rev Immunol 37(1):349–375
13. Athanasiadis A, Rich A, Maas S (2004) Widespread A-to-I RNA editing of Alu-containing mRNAs in the human transcriptome. PLOS Biol 2(12):e391
14. Bazak L et al. (2014) A-to-I RNA editing occurs at over a hundred million genomic sites, located in a majority of human genes. Genome Res 24(3):365–376
15. Kim DD et al. (2004) Widespread RNA editing of embedded alu elements in the human transcriptome. Genome Res 14(9):1719–1725
16. Ku J et al. (2024) Alternative polyadenylation determines the functional landscape of inverted Alu repeats. Mol Cell 84(6):1062–1077.e9
17. Dhir A et al. (2018) Mitochondrial double-stranded RNA triggers antiviral signalling in humans. Nature 560(7717):238–242
18. Szewczyk M et al. (2020) Human REXO2 controls short mitochondrial RNAs generated by mtRNA processing and decay machinery to prevent accumulation of double-stranded RNA. Nucleic Acids Res 48(10):5572–5590
19. Borowski LS et al. (2013) Human mitochondrial RNA decay mediated by PNPase–hSuv3 complex takes place in distinct foci. Nucleic Acids Res 41(2):1223–1240
20. Roulois D et al. (2015) DNA-demethylating agents target colorectal cancer cells by inducing viral mimicry by endogenous transcripts. Cell 162(5):961–973
21. Chiappinelli KB et al. (2015) Inhibiting DNA methylation causes an interferon response in cancer via dsRNA including endogenous retroviruses. Cell 162(5):974–986
22. Katayama S et al. (2005) Antisense transcription in the mammalian transcriptome. Science 309(5740):1564–1566
23. Lapidot M, Pilpel Y (2006) Genome-wide natural antisense transcription: coupling its regulation to its different regulatory mechanisms. EMBO Rep 7(12):1216–1222
24. Shendure J, Church GM (2002) Computational discovery of sense-antisense transcription in the human and mouse genomes. Genome Biol 3(9):research0044
25. Faghihi MA, Wahlestedt C (2009) Regulatory roles of natural antisense transcripts. Nat Rev Mol Cell Biol 10(9):637–643
26. Kim Y et al. (2014) PKR is activated by cellular dsRNAs during mitosis and acts as a mitotic regulator. Genes Dev 28(12):1310–1322
27. Sakurai M et al. (2017) ADAR1 controls apoptosis of stressed cells by inhibiting Staufen1-mediated mRNA decay. Nat Struct Mol Biol 24(6):534–543

28. Elbarbary RA et al. (2013) STAU1 binding 3′ UTR IRAlus complements nuclear retention to protect cells from PKR-mediated translational shutdown. Genes Dev 27(13):1495–1510
29. Williams BR (1999) PKR; a sentinel kinase for cellular stress. Oncogene 18(45):6112–6120
30. Yoneyama M et al. (2004) The RNA helicase RIG-I has an essential function in double-stranded RNA-induced innate antiviral responses. Nature Immunol 5(7):730–737
31. Kato H et al. (2006) Differential roles of MDA5 and RIG-I helicases in the recognition of RNA viruses. Nature 441(7089):101–105
32. Rehwinkel J, Gack MU (2020) RIG-I-like receptors: their regulation and roles in RNA sensing. Nat Rev Immunol 20(9):537–551
33. Stark GR et al. (1998) How cells respond to interferons. Annu Rev Biochem 67(1):227–264
34. Honda K, Takaoka A, Taniguchi T (2006) Type I inteferon gene induction by the interferon regulatory factor family of transcription factors. Immunity 25(3):349–360
35. Crow YJ et al. (2006) Mutations in the gene encoding the 3′-5′ DNA exonuclease TREX1 cause Aicardi-Goutières syndrome at the AGS1 locus. Nature Genet 38(8):917–920
36. Rice GI et al. (2012) Mutations in ADAR1 cause Aicardi-Goutieres syndrome associated with a type I interferon signature. Nature Genet 44(11):1243–1248
37. Zhang Y-J et al. (2019) Heterochromatin anomalies and double-stranded RNA accumulation underlie C9orf72 poly (PR) toxicity. Science 363(6428):eaav2606
38. Saldi TK et al. (2019) Neurodegeneration, heterochromatin, and double-stranded RNA. J Exp Neurosci 13:1179069519830697
39. Ku Y et al. (2021) Noncanonical immune response to the inhibition of DNA methylation by Staufen1 via stabilization of endogenous retrovirus RNAs. Proc Natl Acad Sci U S A 118(13):e2016289118
40. Bonin M et al. (2000) Determination of preferential binding sites for anti-dsRNA antibodies on double-stranded RNA by scanning force microscopy. RNA 6(4):563–570

Subjet Index

Je-Hyun Yoon (ed.), *RNA Labeling*, Methods in Molecular Biology, vol. 3047,
https://doi.org/10.1007/978-1-0716-5352-4,

FSC
www.fsc.org
MIX
Papier | Fördert
gute Waldnutzung
FSC® C083411

Kolibri 360 GmbH
Ferdinand-Jühlke-Straße 7
99095 Erfurt, Deutschland
produktsicherheit@kolibri360.de